Environmental Projection

The Collision of Modern Technology and Sacred Spaces

LUKE McELROY

Published in Nashville, Tennessee through SALT Conferences, a division of Orange Thread Media, LLC.

For bulk, educational or organizational licenses, please contact Orange Thread Media for pricing and details.

Cover Design: Joe Cavazos
Book Layout: Jacob Blaze
Edited By: Rachel Dawn Hayes, Tim Southwick & Victoria Aebischer

For color versions of pictures represented in this book, please visit http://LukeMcElroy.com/EPBook-Bonus. You'll also find a myriad of bonus resources.

Electronic Book ISBN: 978-0-9913307-1-3
Paperback Book ISBN: 978-0-9913307-6-8
Library of Congress Number: 2016903876

CONTENTS

FOREWORD

I didn't know stage design was a thing. When I was working full-time at a church, my job description tasked me with bringing the 50-year-old organization into the modern world. This meant completely rebranding, but also changing the look and feel of our sanctuary. Again, not knowing that stage design was something churches did, I just went with the vision in my head that involved adding some color and pop to the stage. So we did it.

If I'm being honest, my motives weren't the best. I wanted my church to be cool. And in my opinion, chandeliers, wood-beamed ceilings, and shadow-box walls weren't accomplishing that purpose. So we got rid of it all, blacked out the stage, and threw LED lights everywhere. We went from a sacred-looking space to a rock-and-roll show in the span one a month. I wonder if I might have missed out on an opportunity to accent the beauty already in our building, pointing people's eyes toward God. I'm not saying my stage design was bad. It was actually really good. But I missed out because I let "making our church cool" be the driving factor for my design decisions.

Regardless, I decided to post the process on my personal blog. When I noticed that was the most popular post I'd ever written—

by far—I got the idea to start a website about stage design. A few months later, ChurchStageDesignIdeas.com launched. That's when I first met Luke McElroy. He took a risk on me and partnered with me by purchasing an advertisement.

Since then, we've had the chance to work together in many different roles. He's written for me; I've written for him. He's been to the conference I directed; I've been to his. Through all these encounters, I've been impressed with his skill as an artist as well as his heart behind the art. Conversations with him and others in this space have changed my perspective since my first stage design.

In a world of thrilling technology and the ability to tantalize the senses, it's easy to geek out over gear. It's easy to talk about the "how" and forget all about the "why". But, as you'll soon find out, Luke is all about the *why*.

I love that the first few chapters of this book don't discuss technology. It's all about the philosophy behind environmental projection. It's the heart behind all the fun and opportunity of the technology.

I know the eager reader might be tempted to skip straight to the step-by-step instructions included in the book. But I encourage you to wait. You'll get there, and it's as glorious as you're imagining. Before you get into all the technical fun, though, let Luke's thoughts refresh your heart and inform how you can use this amazing technology effectively. Let the heart of visual worship dictate your use of the tools.

Have fun with environmental projection. Experiment with the awesome visuals you can create with the medium. But more important than all of that, use it as a tool to lead your audience into a fascination with our amazing God. Let *this* be a vehicle to *that*. Let the journey lead to new spaces of wonder.

- Jonathan Malm,

author of Created for More, Unwelcome, *and owner of Sunday | Magazine*

INTRODUCTION

Preparing The Meal

I will never forget sitting in a breakout session at the Passion Conference several years ago in Atlanta, GA. Tim Hughes, worship leader and the one who penned *Here I Am to Worship, Beautiful One, Happy Day* and many more recognizable worship songs, was speaking on the process of planning a worship service.

First, let's pause: I'm not a musician and you'll never see me as the front man wearing a V-neck and leading the congregation in the latest Hillsong Young & Free tune. However, I am part of the worship team. My role involves creating a visual atmosphere for people to encounter God.

I was on the edge of my seat soaking up wisdom from Tim as he unwrapped the idea of preparing and partaking in worship.

The room was packed with more than 500 people and I was sitting four rows from the back when Tim started the analogy. He said he was going to invite all of us over to his house for a big meal.

He said he wanted to serve us the most spectacular meal we'd ever had to show us the importance we have in his life. He shared how he would go to the meat market and pick out only the best filet mignon, perfectly marbled and sumptuously tender. He imagined himself selecting a slew of spices and flavorings to add to the meat to bring out a rich, hearty flavor.

He continued to talk about marinating the steaks for hours before cooking them to perfection, searing the outside on high heat and then reducing the temperature to cook the inside to a juicy medium rare, just like Gordon Ramsay does in his five-star restaurants.

To accompany this decadent steak, he wanted to prepare Parmesan-crusted Brussels sprouts charred just enough to add flavor, but avoiding a burnt taste. Also on the plate would be hand-whipped mashed potatoes served with an ice-cream scoop to achieve a perfect round shape as a pedestal for the meat.

As another course, Tim's favorite salad would be prepared with the freshest farm-to-table greens, bold peppers in a variety of colors, rich and plump tomatoes, and an expertly seasoned vinaigrette dressing to top it off.

I remember my stomach started to grumble, enticed by the detail and focus he employed in laying out this meal that meant so much to him.

He finished by describing how he would use the finest china and glassware, spending hours polishing each and every fingerprint off and taking care that the lighting was set to make everything shine and sparkle. He wanted the presentation to be just as rich as the meal and for the decor to bring about an emotional response.

After the table was set and the meal was ready; after what seemed to be days of preparation and fine attention to detail, he would invite us to take a seat at this massive banquet table with him at the head. Then, Tim would invite us to say grace. Once the word "amen" was spoken, we all would open our eyes and with excitement, grab our forks and knives and dig into the delicious meal.

When he finished, you could have heard a pin drop in the room, waiting to see where this story was heading.

He went on to describe how those at the table would eventually notice he didn't have a plate in front of him. One of us would ask, "Tim, are you going to eat?" He would respond, "Oh no, I've been eating all day, this meal was prepared for you, not for me."

What?

Do you see what he did?

Do you understand or comprehend what just happened?

It hit me like a ton of bricks. It spoke to the heart of preparing a space for worship. We spend all this time preparing the environment and the meat of our worship services then don't allow ourselves to partake. We stand up in front of our community or congregation asking them to enjoy a meal or experience that we don't or won't allow ourselves to enjoy.

Wouldn't you think twice about the meal that was being served if the host wasn't eating it? Was there something wrong with the preparation of the food? Was the meat spoiled?

Whenever you and I are involved in the process of preparing the atmosphere of worship, we must be willing and able to partake in the experience alongside our community As we get into this topic a bit, you'll find that the preparation process only increases when you implement technology to create immersive environments. Don't ever let the abundance of preparation rob you from the opportunity to worship.

Regardless of the role you play in preparing the worship atmosphere, we all put together the meal (or the service). Environmental projection may not be the decadent steak or the rich mashed potatoes, rather it's the atmosphere that surrounds the steak and potatoes.

Through this book, I'm going to walk you through the recipe of becoming a master of environmental projection. It's my hope that at the end, you won't merely design a room so everyone else enjoys the feast, but instead you'll be able to design a space where the entire body (of which you play a critical role) will engage on a deeper level with Jesus Christ. Join me as we begin to use modern technology to create powerful sacred spaces.

Definitions

Before we start this journey together, it will be helpful to have a common language to keep us all on the same page.

Over the next few pages we're going to talk about **immersive sacred spaces**. Any room or space that is sacred is designed in a way that allows us to reflect on the character and presence of God while being able to respond in prayer, praise or posture as an act of worship.

For years, church clergy, or staff, have been creating sacred spaces for people to connect with and encounter the presence of God. When we set aside space for prayer and praise (making it sacred) and we add visual elements that surround our congregation (create an immersive environment), it heightens our ability to create an atmosphere in which someone can worship. I call this an immersive sacred space. I believe our whole goal is to create powerful immersive sacred spaces, making space for our communities to encounter God in powerfully relevant ways.

The second phrase we need to have a common understanding of is **environmental projection**. This is a tough concept to explain merely in writing, as it's best to experience its beauty at scale. But imagine projecting on all the walls of your worship center, floor to ceiling, an image of a cathedral or outer space. Imagine your

Church on the Rock, St. Louis, Mo. © 2014 Luke McElroy.

wallpaper becoming a digital canvas of art and beauty. That's the beginning of environmental projection.

It's like the cyclorama you visited as a kid at the local science center with 360° projection. Though usually not that intense, environmental projection is in essence the use of a fully immersive projection surface (usually the walls of your building) to invite people into a greater story.

There are other concepts we will need to have a common language for as we embark on this journey, but we'll cover those as needed. For now, prepare to understand the art of crafting an immersive sacred space with environmental projection.

CHAPTER 1

NOTHING NEW

Visual worship, creative arts and immersive sacred spaces are nothing new. They're not simply a trending topic on Twitter, nor have they been recently introduced due to an uptick in technological advancement.

Atmospheric sacred space wasn't designed to lure in the next generation. Creative arts didn't emerge to make the church appear relevant. Visuals and art were originally used in the early Church as vehicles for God's presence, grace, magnificence, mystery and power to be experienced by communities all across the world. (More of this thought is captured in a video called "The Manifesto" and is available at http://LukeMcElroy.com/EPBook-Bonus.)

Now that you know the terms, "environmental projection" and "immersive environments," we can dig deeper into how these are modern terms for the inclusion of art and visuals used in the worship experience. They encourage us to paint with digital pixels the story of God and invite people into it by offering a setting in which their own story may be seen or experienced.

This book is designed to make a case for immersive sacred spaces and our need for visual environments to be seen not as a distraction, but more as an opportunity to invite people into the grand story God has been writing throughout history. I hope to show you the impact they can make, how to utilize modern technology well, and how to be a part of the rebuilding of the creative arts in your congregation or church community.

As I've said, sacred immersive spaces aren't new. They have existed for centuries. Maybe not in the form of environmental projection or cutting edge stage designs, but the idea that the context of our worship often shapes the content of our theology has been around far longer than the invention of modern projection or LED lighting.

Leaving on a Jet Plane

In November 2012, I got a chance to visit London for the very first time. I was invited to speak at the Mission Worship conference on the topic of environmental projection, and decided that I would add a few days to the front of my trip so I could do some sight-seeing with my dad, who met me in London.

In the 12th and 13th centuries, the Roman Church began building some of the most beautiful cathedrals in all of Church history[1]. Never having set foot in one of these magnificent spaces before traveling to England, I knew a trip to Westminster Abbey or St. Paul's would be a must-have on our trip itinerary.

Long before the terms "environmental projection" or "visual worship" were coined, and long before the idea of a digital screen ever existed, the concept of visual worship was alive and well. These living art pieces, inside St. Paul's or Westminster

Abbey, still exist today. These beautiful buildings, tell a powerfully immersive story.

Taking a look inside St. Paul's Cathedral took my breath away. A beautiful Anglican cathedral, located on the tallest hill in London, was intentionally designed to leave you speechless. Immediately upon entering, I was immersed in the most beautiful visual detail and magnificence imaginable.

Don't forget, this is a church; a gathering place for worship and a powerfully immersive visual environment.

In the 1800s, St. Paul's was actually modified to resemble parts of the dome-like design of another famous cathedral of the 12th century: St. Peter's in Rome[2]. Each of these ancient churches have paintings, sculptures and a variety of other forms of art, to transform the worship space into a visually immersive atmosphere. Proclaiming the story of God on nearly every surface of this stunning

Saint Peter's Basilica dome. © 2015 iStockPhoto.

structure, the space itself (the height, artistic paintings, detailed engravings, etc), renders any visitor into a posture of worship.

The early Church didn't have cutting edge projection technology or presentation software, but they did have the insight that atmosphere mattered. Atmosphere invites us as participants in corporate worship to engage with The Creator through creativity.

A Pattern

St. Paul's and St. Peter's are not alone. I didn't stumble upon the only rare cathedral in the world that was built with such magnificence and grandeur. There are countless cathedrals built hundreds of years ago in Italy, England, Spain and France where art plays a prominent role in the church's telling of the story of God.

From England's Canterbury Cathedral and Chartres Cathedral to Paris' Notre Dame in France, extravagant cathedrals can be found around the world. Though each of their stories is unique, their purpose in the community was aligned.

Cathedrals were built to be the centerpieces of a city or community. They were usually the tallest buildings and were created to be a safe place for both the poor and the wealthy, a commonplace for the people of the community to meet, pray and worship as the body of Christ unified, not separated.

One of the driving reasons for cathedrals' size and grandeur was to provide a reference that the House of God is the storehouse of treasure for all people, not just the wealthy. Cathedrals provided a house, desirable to spend time in, for those who did not have wealth or any income at all[4].

Proverbs 22:2 says "Rich and poor have this in common: The Lord is the Maker of them all[5]." Cathedrals were such an important element in a community. They were the House for the people of God no matter a person's background or position.

A second reason for cathedral splendor is cathedrals were a community effort. God doesn't need us to build His kingdom, but He allows us be involved. The massive works of art, or medieval cathedrals, as we know them today, were incredible collaborative efforts for the community. They showed that by being the Bride of Christ we must trust and rely on the various parts of the body to come together and build the House of God (1 Corinthians 12)[6].

When you look at the roles needed for raising a cathedral the list is quite exhaustive. Master quarrymen, skilled stonecutters, sculptors, mortar makers, masons, carpenters, blacksmiths, roofers, architects, glassmakers, painters, and even electricians in many of the latter built sacred spaces, were all required for success[7].

Third, the great cathedrals and churches of the 12th century took an incredible amount of time to build, making the mere creation of the art in the church not about a single human or individual. Today we're all about crediting those who build iconic buildings and key structures in a major city, because most projects don't take more than a couple of years to complete. However, for God to get the glory in His church, the time it took to plan, build, artfully execute and complete a cathedral spanned well beyond the typical lifetime of an artist. Cathedral makers often didn't live to see their finished work and it was common for the building to be completed by someone else, showing the good hand of our God and his involvement in this process[8].

Visually immersive spaces in our modern worship environments seem to mirror some of the same principles as grand cathedrals. Cathedrals; they are for both the poor in spirit and the rich in faith. Inspired art and sacred space can invite people who are deep in their walk with God as well as those who have just begun. Art enables us to see the world or a story as the Holy Spirit desires us to. Visuals communicate in such a way that words or music can't interfere, as God is the ultimate authority in determining how art moves and speaks to individuals.

These spaces also invite people into the story of God. The best way to make a visually immersive and powerfully artistic atmosphere in worship is to come together as a community of artists, craftsmen and people. Cathedrals weren't built in a day and neither is a worship atmosphere. There is no question that technology has altered this timeline and our ability to build dynamic visually immersive atmospheres, but let's not forget to invite the team into the process and allow community to be a part of the story.

We're beginning to see how the cathedrals of the past impact the worship concepts of today and how visually immersive spaces can affect our communities. It would be a travesty if our modern worship spaces weren't inviting for the poor and wealthy alike, if we excluded any in our communities, or if our spaces were designed for luring a new audience instead of fostering an atmosphere for heavenly adoration and worship.

Finally, the visually immersive spaces we create today are not about us. They weren't in the past, either, as experts are still unsure of who deserves credit for the great cathedrals of the 12th and 13th centuries. Creative arts were never designed to be a social media engagement tactic or something pretty to look at, but rather a powerful opportunity to engage people in the mystery and beauty that is the story of God.

CHAPTER 2

STERILE TO SACRED

Sacred Space

The use of visuals and art in our worship services today has changed from the days of Michelangelo, among the early tribe of artists in the 12th and 13th century. Art and visuals don't play the same role they used to. Can it be blamed on the rapid innovation of technology? Or the church's love-hate relationship with art? Perhaps the role has shifted because our community doesn't see the value of art as they once did. Regardless of the reason behind this change, art and visuals are still vehicles for the Holy Spirit to move people in powerful ways.

As we look at today's modern worship center and walk into a more "typical" (I use this term loosely as I know there isn't a typical church in the world today) Bible Belt church, you'll find yourself in a room that seats 200-500 people[1]. This typical church has seats much like those at the IMAX or local movie theater. The floor will be covered in carpet and the stage is empty so it can be multi-purpose. The walls of the church will be a neutral, light color to provide a common base for any additional accents added during special events.

Our spaces have moved away from sacred and more toward sterile when it comes to the emotion conveyed when walking into the room. The words you would use to describe a space like St. Paul's and the words you would use to describe your own space would be in stark contrast to one another.

When I teach seminars or train local churches on this concept, I'll often show a picture of a cathedral similar to Westminster Abbey or St. Paul's Cathedral and ask my students to describe it. The words I hear include beautiful, reverent, awe-inspiring, majestic, detailed and my favorite, sacred.

Then I show a picture of a modern day sanctuary, one similar to what I described above, and ask the same thing. The words I hear are astonishing. The list most often includes: boring, bland, dead, dry, plain, and the most disturbing of all, sterile.

How has the church moved from "sacred" to "sterile" when it comes to our worship spaces? Whether we acknowledge it or not, our space, our worship environment and our room tell a story to the people who are invited into them.

Every room tells a story. Not just worship spaces, but entertainment venues, theaters, amphitheaters and other rooms where there may be a stage or seats all tell a story of some kind. We'll dive into these in detail later, but color, texture, arrangement and the literal art on the walls tell a story. My fear is that we're not telling the story we want our community to hear and we're not inviting them into the mystery and beauty that is the Gospel.

Is your room telling a story about a God who is magnificent, holy, gracious, full of splendor, beautiful and sacred, or a God who is dead, dry, bland, removed, boring and sterile?

Most likely, your room is telling a story that is far from 'sacred' and much more about a distant, plain, ordinary, 'lower-case g" god that demands we live clean, perfect lives in order to approach him.

The truth is that we don't need to take a trip to London to bring sacredness back to our worship environments. We need a culture that doesn't have to travel the world to see marvels of art in the church when we have certain technological advancements at our fingertips.

It is time for our churches to begin to see our spaces as canvases to tell God stories and not let the unknown of modern tools and equipment scare us from creating powerfully immersive sacred spaces.

A New Paintbrush

A few years ago I hired a full-time artist to join our team. He was the primary creator of movies and videos for our clients. More than that, he was the storyteller for our projects and the lead artist in the branding of our products.

While we're both artists, he and I had different ways of using a computer. For one, I used a mouse and he didn't. I'll never forget the day I had to help him set something up on email and I couldn't use his computer because there was no mouse. He used a different tool, a stylus and tablet. It was like trying to operate the computer with a paintbrush, but one that never ran out of paint or made a big mess.

The same is true with today's technology. It's a new paintbrush for our worship atmospheres. With the advent of new technology we're able to make pretty incredible transformations in our worship environments at the click of a button or stroke of a stylus, bringing the sacred back from the sterile. I have never looked at the "typical" modern church as a sterile environment as I've always seen it as a canvas on which I get to paint a glorious mural.

The canvas isn't new; it's always been there. You have a canvas just like every other church or worship environment. Some spaces feel like true canvases with blank white walls. Some feel like white statues or sculptures with columns. Still others feel like three-dimensional extrusions, which may make the "painting" part a bit more time consuming, but equally as impactful.

Projection, software, content and a little bit of video processing can create powerful environments in a fraction of the time Michelangelo spent on his masterpiece. We get to create atmospheres that beckon people to be in awe and wonder. Often, we as the creative or technical worship leaders get to take a little bit of travel ourselves. It's like we're behind this amazing teleporting machine and with the tap of a space bar or the click of a mouse, we're inside one of the great cathedrals. Another tap and we're dancing with the angels in the heavenly realms.

You and I have been invited to pick up a new type of brush. Throughout this journey we're going to look at what it means to create sacred visual spaces in worship environments. Looking at projection and other technologies of today's culture, we'll see how we can create our own Sistine Chapel. The best part is that your back won't be broken like Michelangelo's was after painting, and you won't have spent your whole life on one project.

Proceed with Caution

As we proceed in learning how to create masterpieces with projectors, I need to give you a strong heed of caution. The brush you're about to pick up is quite powerful. I've seen this paintbrush bring people to such an intimate place with their Creator that the result is tears. More often than not, though, I've seen people walk out of a room due to distraction, frustration, and a deeper anger toward the role of technology in worship.

A good friend of mine visited several years back. I picked her up at the Nashville airport and told her we were going to visit an art gallery. There is one in downtown Nashville that I had never been to and it happened to have an exhibit of abstract art, the kind you can catch yourself staring at for hours upon hours.

After purchasing our tickets and fighting for a place to park, we eventually found our way into the first gallery. It was just as you would expect. The walls were pure white and the dark cherry floor created natural contrast. Walking into the first room we were greeted with hundreds of colorful, bold, abstract paintings.

We continued to explore. In the back corner, by the entrance, was a painting that stood out to me. It was nothing more than a 2'x3' canvas painted with solid black paint. There was no texture to it. It was as if a mischievous child went to town on a beautiful painting, hiding its original message. All I saw was darkness.

For many, I'm sure it had meaning and impact. For me, it was a distraction. I wish I could say I was distracted by the contrast of the painting and its utter simplicity, but that was not the case. I was distracted because I was angry. I was angry because, to me, this "painting" had no right being in an art gallery. This wasn't a painting to me and it definitely wasn't art. It was as if some artist had this grand plan to get his "not-so-creative" work of art into an art gallery by cutting all corners.

Fortunately, there was another room just around the corner. I was able to walk past the black painting to other beautiful paintings.

This next room was very different from the previous room. It was barren. The walls to the left and right of the entrance were empty, but we could tell that there was another room to the left only a few steps away. As we turned this second corner I noticed one of the largest paintings I've ever seen with my own eyes. It covered the entire wall.

It was the only painting in the room—a beautiful abstract painting of deep oranges, pinks, ambers and the slightest stroke of red. What took me aback wasn't the fact that it was the focal point of the room it WAS the room.

Then it hit me.

I started thinking about that odd black canvas in the other room and compared it to the majestic scene before my eyes. And then I realized how similar this was to immersive projection and our worship environments.

Whenever we use media in our worship services we're showcasing a canvas before our congregation. If that media lives on a 6'x8' screen on the side of our stage then it's like having that 2'x3' canvas in an art gallery. If it's distracting it can only do so much damage because it's a small part of the grand design and visual story of your entire room. Creating an immersive visual environment through environmental projection is very different, as the entire room becomes the canvas of the art you're displaying.

When your media stretches from one side of your room to another, it becomes the only canvas in the room. It's not just the focal point, it's the room itself. The room begins to be centered around this canvas and you begin to feel the environment that the art is creating.

I would have been furious if I had walked into that second gallery and the massive painting was solid black. It would have been such a distraction and one that I wouldn't have been able to just pass over to continue marveling at the beauty of other pieces. The painting was the room, which means the room itself would have been a distraction.

The canvas of immersive visual environments is massive, therefore we must proceed with caution to be sure the impact of our environment is one of a positive nature that adds depth and meaning rather than noise and distraction.

Harry Potter

Now that we have realized the incredible distraction that visual environments, and more specifically, environmental projection, can bring to any room or worship service, we can now talk about the incredible impact it can have on our worship services.

We've all heard the statement, "A picture is worth a thousand words". Where did that come from? The answer is from a newspaper editor in 1911[2]. It's not a statement that any of us are confused by. In fact, I find it ironic that a newspaper editor said it. How often do you see pictures in a newspaper article? Not as often as you would think if a senior editor said, "a picture is worth a thousand words."

Do you think the 12th century church realized the impact of pictures?

When you walk into those beautiful cathedrals the rooms are wrapped in visual images and pictures. There are few mediums of communication that have the depth an image has. If we accept this as fact, it means the way in which we process information from a picture is one of the most effective ways to collect information. Let me paint this another way.

If I were to give you a week to provide for me a detailed report of the storyline in the Harry Potter series, you would have two choices. First would be to read all seven books, which according to

Scholastic would be over 1 million words and approximately 3,400 pages[3]. If we were to multiply that out by the average reading speed of an educated person (1/2 page per minute) it would take 6,800 minutes[4]. That's 113 hours and 20 minutes of straight reading. Yes, it's possible without sleep to do that in one week.

However, the second option would be to watch the eight movies (the last book was split into two movies). If we were to take each of the movies and add them all together we would get 1,178 minutes of film[5]. That is only 19 hours and 38 minutes. A fraction of the time it would take to read the books.

For avid readers out there who have read every single word in the million-word series, let me make a comment: I do not believe that words and pictures are fully interchangeable in the way in which they communicate. There's no question that the Harry Potter books are much more informative and detailed than the movie series. But, the point remains that an audience is able to experience the Harry Potter story on film in much less time than by reading the book series.

Shane Hipps makes a great illustration on the power of an image in his book *Flickering Pixels*[6]. He is talking about the fundamental differences between visual and textual communication. He shares with his reader the sentence "The boy is sad." On the next page he shows a captivating picture of a boy with a tear streaming from his eye. Without reading anything you've immediately connected with this boy. Shane made a great point with the following conclusion:

> Images initially make us feel rather than think. They can pin the logical side of your brain to the back of your skull, which is why image-based advertising is so effective. Images don't invite you to argue; they give you an experience. In contrast, the painted words make us think and question. The statement "The boy is sad"

generates several logical questions: Which boy? How sad? Why? But one look at the image leaves no doubt of the boy's sadness.[6]

When it comes to our worship services we can use images to help convey feelings and emotional connections to our Heavenly Father. Imagery enables us to spotlight that text or song on an intimate level. When the story aligns (i.e. the music, text, pictures and the Holy Spirit) we have then created a catalytic environment that lays the groundwork for life change.

What to Expect

I hope this book is a launching pad for your visual worship journey. More specifically, I hope it encourages you to expand your knowledge and grows your confidence to create visual environments as a way to usher in a powerful atmosphere in your worship services.

Over the next several pages, I hope to shed some light on the mystery of environmental projection and immersive sacred spaces. My prayer is that I'm able to extend to you the "secrets" of how to use this technology. I plan on sharing with you all the details of environmental projection, including the elements needed to create an immersive environment, how to set it up, tips on finding great media that tells a narrative, and how to begin the process of creating a visual "style sheet" for your community and context.

If you're ready for an adventure through the world of projection types, media curation, and the alignment of projection to transfer people to places they've never before experienced, then continue reading. If the idea of telling a story with your space gets you excited, then welcome to the journey.

Welcome to ***Environmental Projection: The Collision of Modern Technology and Sacred Spaces***.

CHAPTER 3

THE MANIFESTO

In The Beginning

When I first started my very own visual worship journey I had no clue what I was doing. To be really honest, I had the whole entertainment and spiritual depth thing mixed up. There was no question that I was choosing visuals and motion backgrounds for one purpose: it looked cool on the screen!

I grew up in a fairly large church environment and we always seemed to have new tools and technology around us to play and experiment with. I even remember dreaming up an opener for the worship service when I was in high school that included the theme song to Batman, because we had recently bought Batman gobos (or cutouts to make the light look like the shape of the batman logo) for our moving lights. We were so excited about this because Six Flags over Georgia had just released *Batman: The Ride,* and we wanted to emulate that (as the opening to our worship!). To say we had it all backwards would have been an understatement.

The greatest blessing during this season was my church's hiring of a new technical director. His name was Steve and he was incredibly knowledgeable in all things technical. Even better, Steve happened to be an ordained minister. This man played a key role in helping me understand the power of media and technology in a worship service, all the while teaching me what it meant to be a disciple of Christ. His ideas and philosophies encouraged me to begin digging into the Word to learn about worship, particularly, visual worship.

I cannot imagine where I would be on my visual worship journey if it weren't for Steve unlocking this Pandora's box and showing me how technology and theology go hand in hand rather than being on opposite ends of a spectrum. I find when I meet people across the nation they feel their gifts and their beliefs weren't meant to collide, as if the faders on a soundboard can't actually be divinely inspired.

My interactions with Steve, his helping me see the power between technology and theology, sparked in me a burning passion to know what the Bible said about the role of visuals and technological advancements in worship. I wasted no time at all and went to the place all desperate men of God go to: Google. Entering the phrase "projection in the Bible" was fruitless. I changed my search term to "motion backgrounds in the Bible" to be a bit more broad ... still no luck.

I was frustrated, but God reminded me that He didn't want me to just have the answer but instead to go on a journey with Him to seek His truth. As crazy as it was, I was convinced that as I read the Bible, God would give me some insight on whether or not visual worship was something Biblical and something He cared a good deal about. I had no idea what I was about to discover.

Bob and Larry

I started in Genesis. I was hoping God would open my eyes to something I had never before seen. Maybe something very specific about visual worship, maybe something true about Him that I could relate to in creating media, projection, or for that matter, environmental projection. I eventually made my way to Exodus.

This was the book that changed my life, but it wasn't until the end of the book. Exodus is the story of Moses and the exodus of God's people out of Egypt and into the Promised Land. My favorite part is in Chapter 34 when Moses comes down from the mountain:

> When Moses came down from Mount Sinai, with the two tablets of the testimony in his hand as he came down from the mountain, Moses did not know that the skin of his face shone because he had been talking with God. Aaron and all the people of Israel saw Moses, and behold, the skin of his face shone, and they were afraid to come near him. But Moses called to them, and Aaron and all the leaders of the congregation returned to him, and Moses talked with them. Afterward all the people of Israel came near, and he commanded them all that the Lord had spoken with him in Mount Sinai[1].

What a beautiful picture! God's glory was so evident on Moses' face that it was radiant! I love that. (An aside, is God's glory so evident on your face that it radiates for others?) Moses comes down off the mountain after spending a considerable amount of time with God, during which God gives Moses a whole slew of directions on how to lead His people.

Starting in Chapter 34 and ending in Chapter 40, Moses begins describing their conversation. Some find this section of Exodus to be incredibly boring and stale, but I found it to be a world of enlightenment because five of the last seven chapters in Exodus (all describing Moses' conversation with God) deal with the visuals of God's worship.

Moses goes into incredible detail about the texture, color, fabric, angles, cuts, design, craftsmanship, yarn, weavings, etc. to be used in the worship environment. Don't believe me? Go and read the last half of Exodus! When he gets into the Arc of the Covenant, the Tabernacle, the Altar of Incense, the table and so on, the detail is astonishing. There is so much focus on color and material, it really amazes me.

However, Moses begins the entire section by calling forward two people: Bezalel and Oholiab:

> Then Moses said to the people of Israel, "See, the Lord has called by name Bezalel the son of Uri, son of Hur, of the tribe of Judah; and he has filled him with the Spirit of God, with skill, with intelligence, with knowledge, and with all craftsmanship, to devise artistic designs, to work in gold and silver and bronze, in cutting stones for setting, and in carving wood, for work in every skilled craft. And he has inspired him to teach, both him and Oholiab the son of Ahisamach of the tribe of Dan. He has filled them with skill to do every sort of work done by an engraver or by a designer or by an embroiderer in blue and purple and scarlet yarns and fine twined linen, or by a weaver—by any sort of workman or skilled designer[1].

I don't really know how to say these two guys' names, so I just call them Bob and Larry. I love that God, through Moses, asks Bob and Larry to be the "creative directors" of this project. They were given every skill needed in order to make all the elements of a worship service Moses had outlined for them in incredible details. And then it hit me. I am Bob! I am Larry! God called me, at that specific time, to be a creative director and pursue Him in the process of leading the visual design of HIS worship environment.

You're Bob... and You're Larry. And it's time for your pursuit of visual worship to begin. God cares a great deal about the visual of His worship and He's given you the gifts and skill set to lead this endeavor.

Marketing Multiscreen

Whenever new technology surfaces and we are presented with an opportunity to include it in the worship space, we have to explore the medium to determine its value and its place. For environmental projection, there is a fine line to be drawn between entertainment and spiritual impact, as I briefly mentioned earlier in this chapter.

There's no question new technology can be distracting.

There are countless people who have challenged the role of environmental projection in the church. I agree with some of them. It's not for every community. Think about hymns for a second. They work in some church communities, but in others (maybe where the community is formed mostly of those burned by the corporate church) they don't lead to anything good. You can stand on one side or the other, but ultimately, it's just a different form or expression of corporate prayer.

Environmental projection is really just an advanced form of multiscreen. As I outlined in my first book, *The Wide Guide: Blueprint for the Multiscreen Movement*[2], (As a bonus to purchasing this book, you can download a free copy of *The Wide Guide* at http://LukeMcElroy.com/EPBook-Bonus) there are a variety of multiscreen designs. It may be helpful to understand what I mean by the term "multiscreen" before we get too deep into this topic.

Whenever you have more than one video surface, or video screen, combined to create one seamless image, that's multiscreen. Instead of having several televisions side by side with the same image on each, I mean having several televisions side by side where they all combine to create one image.

You've probably seen multiscreen in more places than you can recall. The most common place is in a movie theatre or a fast food restaurant where the menu is displayed across multiple televisions. This is multiscreen, one image displayed across multiple surfaces.

Another example may be the arrivals and departures signage at an airport. Though it's not as fancy as the multiscreen I'm going to talk about in this book, it's still a set of data that is using more than one display to convey information. Imagine putting every arrival or departure from Atlanta's Hartsfield International Airport on one small television. No one would ever find their flights or make their connections.

Just like most elements in the church, especially new technology, there is a fine line between distraction and impact. However, multiscreen shouldn't be considered a mere marketing gimmick for the next generation. Just because MTV and American Idol use multiscreen as a key portion of their stage set designs isn't a good enough reason to include these forms of technology in our worship environments. If so, we're chasing entertainment, and entertainment won't change lives for eternity.

Multiscreen is a form of storytelling that transcends auditory noise and touches the heart with an engaging visual environment of beauty, truth and majesty.

National Worship Leader Conference - 200-foot wide environmental projection setup. © 2013 Luke McElroy

A Story For The World

Multiscreen doesn't have a specific recipe. Twenty-seven separate video screens all connected together could be multiscreen. A 200-foot-wide projection setup with three projectors is multiscreen. Setting up a multiscreen system can be simple (in the case of three projectors in environmental projection) or very complex (in the case of the camp picture with twenty seven televisions). No matter what the setup is, each screen operates as a piece of a bigger puzzle to display one singular image.

If I can venture off of multiscreen technology for a second and bring it back to the spiritual side, you'll see why this is important. You and I are multiscreen in the grand story of God. The story we are telling and sharing with the world is bigger than the high definition screen we watch the news on every night. We're telling a life-altering, ground-shaking, monumental story. We, as a community of believers, are telling one singular, massive story across the various canvases in which we live.

When we utilize technology in a way that amplifies our philosophical and theological beliefs, everything in our services has the opportunity to go deeper. We begin to use technology and media not out of ritual, but out of a deeper and more profound understanding of who we are in light of God. We begin to use art to illuminate our own inner worship.

Big Stuf Camps - 27 televisions used as a multiscreen canvas. © 2010 Tyler Hirth.

When we use environmental protection or multiscreen technology, we get to intersect the contagiousness of the Gospel message with the power of visual media. This intersection breeds an experience. These experiences can transform the way we look at God or the way we interact with the message, transforming our hearts and minds in ways words simply cannot.

This experience could be as simple as letting all the names of God show up on the walls of your worship space during a powerful anthem of praise (as you can see in the picture below where my friend and originator of Environmental Projection, Camron Ware, used a piece of media he created to fill the atmosphere of our worship environment with the various names of God). Or, it can be as complex as using projection as the digital set piece, showing the transformation in the heavens during the few minutes after Christ's death on Golgotha.

Seminars4Worship "Names of God". © 2010 Luke McElroy.

Will You Love Me in the Winter?

Regardless of the application or setup, we're trying to penetrate every aspect of our service with art. This art has the opportunity to speak louder than any word, song or phrase. Multiscreen, in a sense, invites you into the story rather than just listening to the story. It replaces hearing only with feeling and connecting to the story.

I remember a few years back working an event for Integrity Music called *Seminars4Worship*. It was a regional worship conference that I was asked to speak at and assist with production elements. Stephen Proctor, a curator of visual worship and a great

"Will You Love Me In The Winter" © 2010. Luke McElroy.

friend was there, as well. He was responsible for creating the visual atmosphere for the conference through environmental projection.

One of the worship leaders, Brian Doerksen, had just written a powerful worship song and wanted to sing it over the conference attendees. The song was "Will You Love Me in The Winter?" The song asks God if He will love us even when life is dark, dreary and desperate. Asking if God will still remember us when we're far from the joy of spring and we've failed. The first time Brian sang this song at the multi-city event, Stephen didn't do anything visual as Brian designed it to be a moment for worship leaders to reflect without distraction, but everyone fell asleep. Not the type of reflection Brian was hoping for with this deep, somber sonata.

Stephen had an idea. Instead of visual silence, he wanted to create an atmosphere that would help express the feeling of the song and try to engage the heart along with the mind, allowing for a much deeper reflection. So we did just that. Stephen took a few clips of snow falling, which he had curated over the years, and a few images of a thick forest of snow-covered trees, and he created an atmosphere.

Stephen's idea worked! Instead of people checking out, the visuals allowed them to connect to the song on a much deeper level. They were invited to feel the song and connect with the message. This moment quickly became one of the more meaningful moments during the conference and was exactly what Brian was hoping for, deep reflection.

The projection didn't take away from the desire of the worship leader, it added to it!

A Theology of Visuals

As I mentioned in the first chapter, visual worship has been around for centuries. As we have uncovered, a click of a button can do well more than just "paint the walls." It can transform the very philosophical representation of our worship space and reflect our spiritual theology.

If you're tracking with me, then what I am proposing is that with environmental projection we can bring the cathedral home to our own communities. We can go from a celebratory worship moment with tons of energy and excitement to a sacred and reverent worship song. The high-energy motion of particles and light is able to transform into a cathedral image in no time at all and we're able to take people on a visual journey instead of an auditory one only.

We're not telling two different stories, but one collective story on multiple canvases. That's multiscreen. It's a changing theology of visuals that aren't just backgrounds to a boring font displaying lyrics. It's the contagious and rich use of visuals and atmosphere.

I hope I've been able to get you excited about the heart and purpose behind environmental projection, because at this point, I'm going to talk more in depth about what goes into environmental projection and how to set it up. I hope to help you become an expert at the technology so you can create a visual transformation for your community and enrich the worship experience.

Let's get started!

CHAPTER 4

THE EVOLUTION OF STAGE DESIGN

Over time, an evolution has occurred in church stage design and church stage layout. In fact, the difference is so drastic that the ancient church probably wouldn't use the phrase "stage design" or "stage layout." I've made a few diagrams (that are on the following pages) in order to illustrate my point.

The Traditional Stage Design

When looking at the layout of the traditional church you will notice that the pulpits are on the extreme sides of the stage, the choir is on the sides looking toward each other instead of toward the audience, and the altar and cross are the focal point (the shaded oval shows the focal point in each diagram). There is deep meaning and reasoning behind all of this.

The more traditional model of choir stalls (formal name) facing one

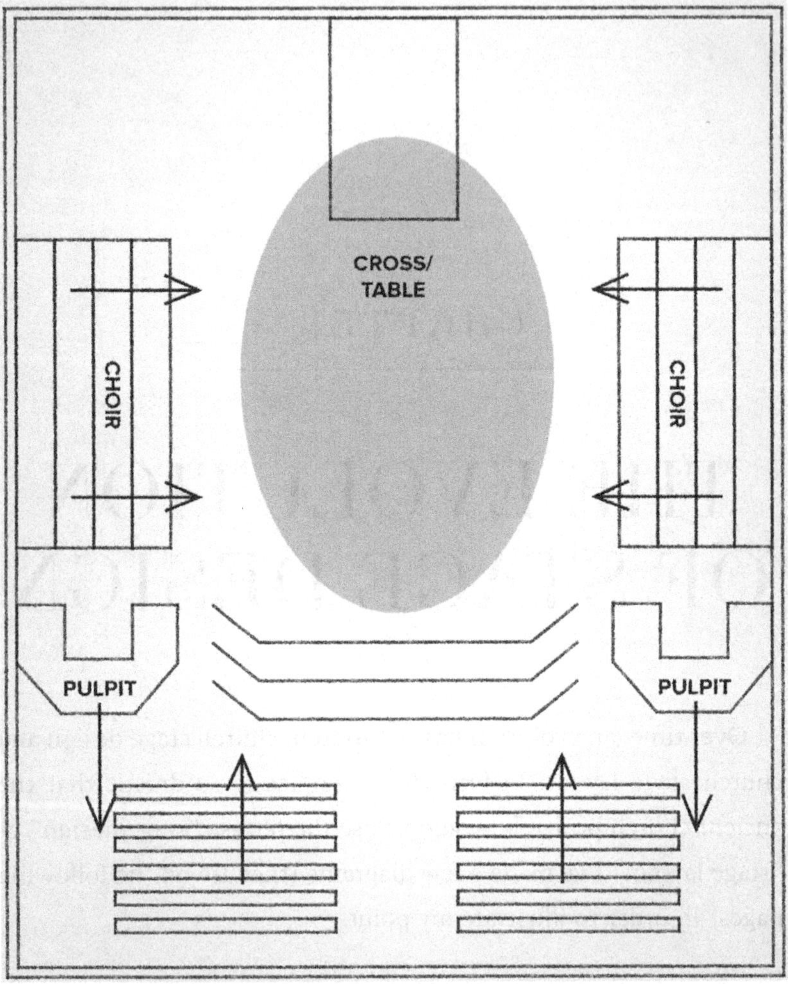

Traditional Stage Design. © 2016 Luke McElroy.

another was to create a sense of centrality for the altar and the cross. Therefore, no one is the focal point. No person is the center. Think about it, the pastor is off to the side. There's something beautiful about that.

The Modern Stage Design

When looking at the layout of the modern church you'll notice

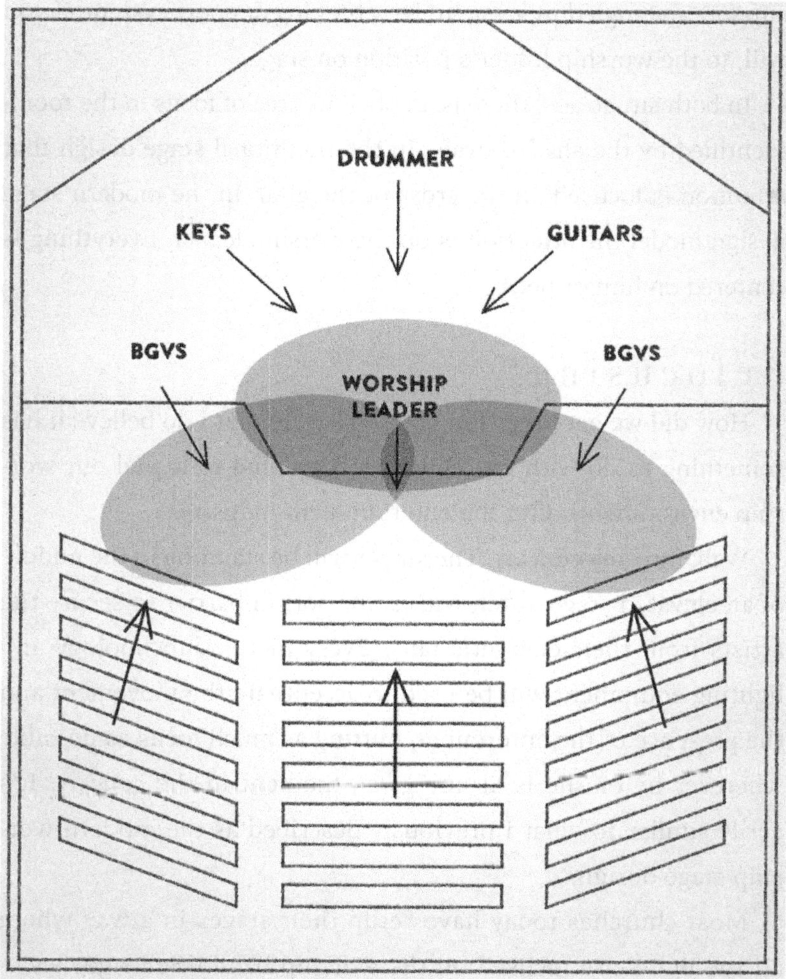

Modern Stage Design. © 2016 Luke McElroy.

quite a few differences. First of all, the worship leader is central Notice how the ovals overlap the Worship Leader position, because there's a lot of visual focus on him or her). The audience is turned in a manner that reflects a posture toward the center (maybe something the traditional church failed at) and there is a band behind the worship leader. The choir (if there is one) and all instrumental support are angled in a way to be a backing support system, if you will, to the worship leader's position on stage.

In both situations, there is an obvious area of focus in the room, identified by the shaded ovals. In the traditional stage design that attention is focused on the cross or the altar. In the modern stage design model the attention is on the worship leader. Everything is centered on him or her.

Refocusing

How did we get here? I'm not 100% sure, but I do believe it has something to do with our churches beginning to model our worship environments after the entertainment industry.

Walk into any concert. The singer will be standing in the middle of an elevated stage, often with some sort of barrier to secure the artists from their energetic fans. Every piece of technology and lighting equipment will be used to accentuate the movement and the presence of the entertainer, putting as much focus as possible wherever he or she is at any given moment in the concert. It's eerily similar to what I previously described as the modern worship stage design.

Most churches today have setup their stages in a way where all the lights are focused on the worship band. We've got backlighting that seem to accentuate the energy the band provides. There's an elevated platform and sometimes an altar wall is present, providing a barrier we may have never noticed before. Over the years, we've incorporated cameras in the room to emphasize the actions of our worship team. The screens are bigger and bigger, the text smaller and smaller, and we wonder why it feels like sometimes we're watching worship, rather than being invited to participate in worship!

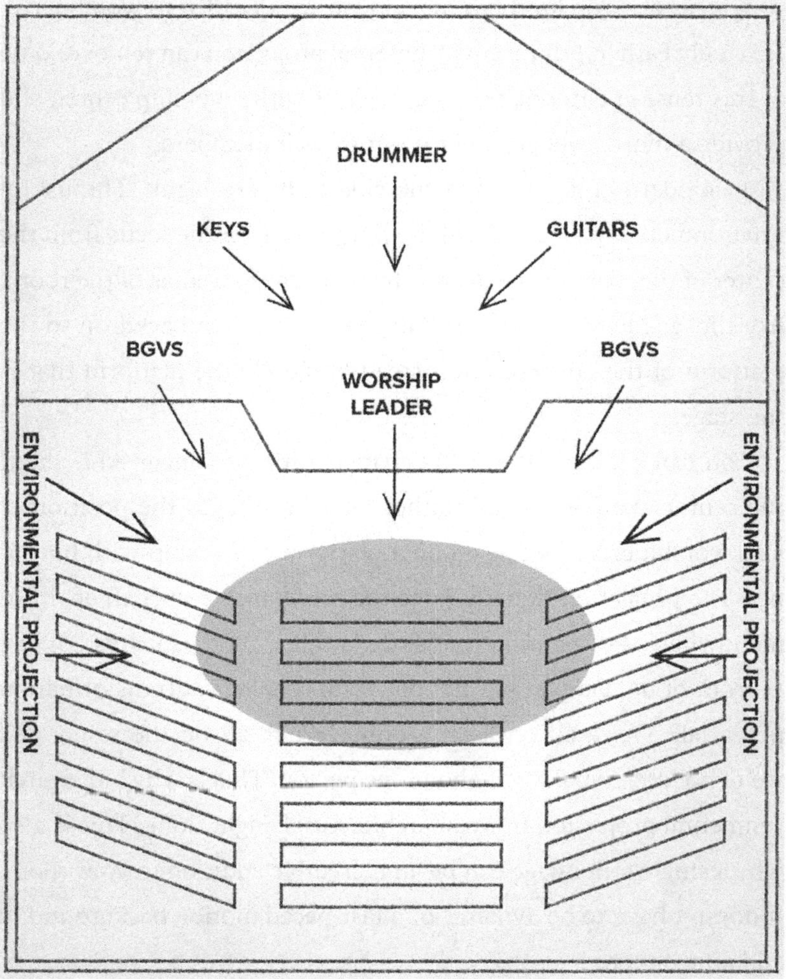

Environmental Projection with Modern Stage Design. © 2016 Luke McElroy.

You probably see where I am going with this. We have somehow transitioned from creating a worship environment where the cross or the communion table is the focus, to a worship environment where a person is the focus. We come to watch someone worship on a platform. This may eventually affect how we engage the church body in worship. The truth is, as the body of Christ, our entire congregation is on level ground and we are all called to worship our Heavenly Father. I think environmental projection can remove some of this tension without re-designing our entire worship center, and provide a more level ground for our church members.

In modern church design the side walls are blank. The use of Environmental projection will help take some of the focus from the center of the worship stage and shift it to the extremes of the room, like those side walls. Now the images serves as a backdrop to the platform of the entire room instead of merely the platform that is the stage.

You can see how the focus changes. It's no longer ALL about the center worship leader. Rather, he or she takes the position of lead worshiper and we're all on a platform of worship with him or her. The projection literally hugs the room and creates an environment where the "lighting" or "stage" is all around us instead of just in front of us. Do you see the powerful theological transformation potential? We're all on level ground. We're all on the stage, and we're no longer making it about one person. This is why I love environmental projection from an architectural angle alone. This is also why a single still image can be an incredible addition to your room. It doesn't have to be dynamic or a fast-paced motion background to add atmosphere.

CHAPTER 5

INTRODUCING THE FOUR INGREDIENTS

Just Add Water

My brother got married in 2009 and some of his groomsmen planned a trip to St. Augustine, Florida, for a bachelor party of sorts. Imagine 20 college guys crammed into a beach house for an entire weekend! It was an experience, that's for sure. There's one thing that's true about college students … they don't have money. That was the case with our group too. We got very creative about using paper towels as plates for most meals, Crystal Light lemonade for our thirst fix, and pancakes for breakfast.

These pancakes were unlike any pancakes I had growing up as a McElroy. The directions on how to make them were astonishing. They said the following: "Just add water."

More amazing than the instructions was the fact that they tasted, well, like pancakes! I'm sure a true pancake connoisseur would have been able to tell you all about the inconsistencies of the batter, and the absence of a real egg, rich whole milk and the slab of butter that makes a true pancake mix.

Conversely, I saw a guy on television just the other day make pancakes differently. He claimed they were "gourmet" and said it was more like an art form than a mathematical formula. He would toss a few ingredients into a bowl, taste it, then add a few more of the missing ingredients his palate seemed to catch, fine tuning his recipe. He proceeded to take his mix and fill the bottom of a skillet (which he made sure to share had been used for generations, thus adding even more flavor). After the pan bottom was entirely covered, he emptied it back out so that there was just a thin layer of pancake mix.

After only a few seconds on the stove, he carefully flipped the thin pancake and cooked the other side. The odd thing was he never called it a "pancake" but rather a "crepe." He didn't want it to be confused with something with simple instructions of "just add water".

Making Crepes

When it comes to environmental projection we can't just add water to some step-by-step tutorial online and call it a day. It's not that easy. We have to learn how to make crepes. Don't worry, though, that's where this book comes into play.

Whenever I teach a seminar on environmental projection at a conference, I get asked questions that can't be answered because people are trying to make just add water pancakes and not crepes. Every single one of you reading this will need a slightly different recipe. I know this because no two churches are alike. God didn't make different church congregations (or you for that matter) to be alike, but to have unique purpose in the world and communities in which they exist.

So let's begin to make crepes together. Get ready to fine-tune your recipe for your community's taste.

When it comes to environmental projection there are four main ingredients that go into every setup. Those four ingredients are software, hardware, processor and content. But remember this isn't a mathematical formula, but art.

CHAPTER 6

INGREDIENT ONE: SOFTWARE

The very first ingredient in environmental projection is a software that will allow you to orchestrate the "environment" in environmental projection. If you don't have a way to generate an image, it's going to be really tough to use images to create a visual environment.

There are a few presentation software options for environmental projection. You may need a simple product that just puts texture up on the wall or a solution that provides advanced control over where the image displays. (Or more importantly, where the image doesn't display. A topic we'll cover later.)

Software is only as good as the computer it's installed on. For any software, you will want a computer that has enough processing speed and memory to handle it. For example, if you have any desire to use moving backgrounds or video in your environmental projection setup, you may want to consider a more top-of-the-line computer with an enhanced graphics card, increased ram or an updated computer processor. Environmental projection deals with large resolutions (we'll get into the specifics later but this translates to larger file sizes and considerably more pixels), invest in an upgraded system to lower the chance of a hiccup or stutter in video playback.

Most software will outline the minimum system requirements in its documentation. Though this is a good benchmark to go by as you plan your equipment purchase, I would highly recommend not purchasing just the bare minimum. Environmental projection is an application that requires a significant number of pixels, and therefore, both the media that you are displaying and the output size of the presentation software will be substantially larger than what the minimum requirements were designed to process. Often, you'll be producing an output three to four times the width (in total pixel count) of a traditional single screen or monitor. Purchase a computer that has a very good processor and outputting graphics card to ensure smooth playback and clean transitions.

There are a few different types of software available, so I've broken this first ingredient into three basic categories:

Single Window

Single window is basic presentation software and the least expensive option. Included in this category is Microsoft PowerPoint, Apple Keynote or OpenOffice.org's Impress. You're probably familiar with at least one of these applications. They all allow for the output of an image, but provide limited control over what can be done with that output in a live setting. Single window applications are perfect for still images. Their features may be limited, but they will allow you to walk before you run and to test your presentation recipe without investing too much money in more complex software.

Dual Window

The next tier of software is dual window. These programs are a bit more advanced (and also a bit more expensive) than single window, but are well worth the investment. Examples include ProPresenter, EasyWorship, MediaShout, Proclaim Online and a handful of other options. I refer to these as dual window because they display a control window on your computer screen that is separate from the output view.

The power of dual window software allows images to be adjusted after a service or event has started. You can quickly get back to that place in the song lyrics without the congregation knowing the difference.

There are many unique features in each of the examples I listed above. As well, check out Articles.TripleWideMedia.com for more in-depth reviews of dual-window software. Regardless of the differences, each of the software options in this category is great for seamlessly looping video files.

To give you an idea of pricing, you'll likely spend around $400 for a good piece of software in the dual window category. This is the most popular category of software for environmental projection.

Multi-Layer Mixing

Multi-layer mixing is the most advanced category of software, and not for the faint of heart or novice user. This category is unique in that it allows you to play multiple videos at the same time and on top of one another. Examples include Arkaos GrandVJ, ProVideoPlayer 2, VDMX and many, many more. Most of these software options allow advanced pixel mapping, that is, complex video setups over multiple video surfaces (say 27 televisions or 7 screens). Pixel mapping is just a fancy phrase for moving specific areas of a video to a specific screen or area of the projection surface, like a collage of different shaped squares.

Some call this category "VJ Software" or "advanced multiscreen." Pricing varies greatly from software to software in this category, from $300 to $1,000+ in certain applications. Though it is possible with some hacks, these applications do not have a built-in text generator. This means you can't just add lyrics to a slide or image.

Let's say, for example, for Easter you wanted to take a still image of an empty tomb and use it in environmental projection. In order to build energy, when the chorus hits, slowly add a motion background of clouds moving on top of the image. With multi-layer mixing software this would be possible. There's a bit of a learning curve to this type of software. There are two main things to learn, the tool itself, and the art of using the tool.

If you wanted to learn how to play guitar like Jimi Hendrix, you would first have to learn how to play all the notes on the Guitar. After mastering the notes, then you would have to learn how to accomplish the particular tone in each of those notes to match those of Jimi Hendrix.

With any of these multi-layer mixing software, you'll need to understand the interface, the way various layers mix with others, effect channels, and what each specific feature can do to your media. Once you have a good grasp on what the potential is and how to use each feature to accomplish a specific look, then you'll need hours of practice to get the timing, tempo and visual "tone" mastered.

This software category is a lot like a musical instrument, where you have all the control in the world to affect virtually every element of a motion background or visual output.

CHAPTER 7

INGREDIENT TWO: PROJECTOR

Now that you have a computer and software to play back a video or image, you need a way to get that video or image onto the wall or projection surface. You need a projector. When you factor in the variety of features—brightness, lens options, input types and shape of each projector—the right projector could result in hundreds, if not thousands of options. Rather then outlining the exact projector make and model you need to purchase, I will break down the technical lingo and simplify the process of choosing the right projector.

When it comes to projection, follow this rule of thumb: **Your goal is to be paint on the walls, not a premiere broadcast of the weekend football game!** This helps you decide if it's worth going high definition (HD) or standard definition (SD). You don't need to see the water molecules on the tiny blades of grass, so you can usually get away with SD.

This rule also applies to brightness. When considering brightness, you want to make sure the output is bright enough to impact the people who will be moved by it, but not so bright it would distract those whom this wouldn't impact. As I mentioned before, anytime you incorporate new technology into a worship environment, there's likely going to be a group of people distracted by it's inclusion. Therefore, we want to make sure that the brightness of the images being projected onto the walls of our sanctuary aren't too bright for people to be able to tune them out. If this group of people closed their eyes but they could still sense movement of our projected images, then our projectors are likely too bright. However, we don't want the images so dim that they virtually loose their impact.

That said, let's walk through the various attributes of a projector, as finding the right one can be an overwhelming task:

Projector Technology

First, let's examine the various projection technologies available—that is, how a projector actually creates an image. There are pros and cons to each technology.

LCD (Liquid Crystal Display): These are the most popular and often the most inexpensive projectors on the market today. The technology uses three crystal panels, (also know as "chips") which give this technology its name by creating an image out of these liquid display panels made of crystal. As the images are processed, the projector allows various amounts of light through these crystal panels. One panel is specifically for the amount of green light passed through, another for blue light and the last one for red light. By allowing light variations to pass through these different pieces of glass, the projector is able to create the image we want.

To make more sense out of this technology, let's go back to overhead projectors. If you could only print in red, green and blue ink and I asked you to put a purple square on the screen, you would have to print a solid red square on one transparency and a solid blue on another to do so. When overlapping these two transparencies, your projected image would be purple. That's essentially how LCD works. It's the overlapping of three "transparencies" of color in order to make up your desired image. It is a slightly older technology, which is part of the reason it's cheaper. The simplest way to explain LCD technology is this: Light passes through a chip in order to project an image.

The Pros to LCD Projectors:

- Generally more cost effective
- Low maintenance cost
- Better access to replacement parts
- More portable
- More varieties; more options to choose from during purchase process

The Cons to LCD Projectors:

- Low contrast ratio, tough to achieve true black
- Shorter lamp life
- Images can be "burned in" if they use too much of one color too long
- Increased chance of "dead" pixels
- Honeycombing effect more apparent (when you project really large, you'll see your image seems to have a grid laid on top of it. This is the separation of various pixels on your chips.)

- May need to re-calibrate the pixels more often through a complex process of convergence, which requires specific tools

DLP (Digital Light Projection): This technology actually has a circus going on inside of it. Not really, but there is a light source (just like LCD has) aimed at a wall of micro mirrors that reflect light in various patterns. To create white, the mirrors move to allow light to pass through. To create black, they move to reflect light away from the lens. All this movement produces a black and white image. (It's like a flat disco ball). To create color, a wheel of colored glass in front of the light source changes color as needed before it passes into the mirrors. Sounds like a circus doesn't it? A circus costs a bit more than an overhead transparency; therefore, this technology is pricier. Put simply, DLP technology is light reflecting off a mirror in order to project an image.

The Pros to DLP Projectors:

- Better contrast (whiter whites and blacker blacks)
- Better color brightness (on the 3 Chip DLP, not true for single-chip DLP)
- More uniform colors throughout image
- Less hot spot from lamp
- Lower chance of a burned in image since it's entirely digital

The Cons to DLP Projectors:

- More expensive (generally speaking of 3 chip vs LCD)
- Tough to punch "brightness" beyond standard settings
- Noisier operation due to spinning wheels
- Can provide "rolling effect" of image when using certain video cameras due to refresh rates and spinning chip set
- Inconsistent quality across brands

Brightness

"How bright of a projector should I get?" Remember this isn't a mathematical formula, it's a work of art!

First, you should know that projector brightness is measured in lumens (technically, it's called ANSI lumens). ANSI Lumens is a unit of measurement created to attempt comparability across various projectors. However, not all projectors with 3,000 lumens are identical, but the number of lumens should give you a rough benchmark on brightness. Simply put, the more "lumens" a projector has, the brighter a projector is. Therefore, a 5,000 lumens projector is going to be significantly brighter than a 3,000 lumens projector. Projectors range from 1,000 lumens to 40,000 lumens. (You don't need that bright of a projector, and you probably won't want to spent the $125,000+ on the 40,000 lumens either). Most medium body (or medium sized) projectors are 3,000-10,000 lumens, and will fit most worship environments.

There are a few things to keep in mind when determining what projector lumens is right for you:

1. **Ambient light.** Ambient light is the amount of unintentional light in the room. This would include a reflection of light from your stage (if unable to control), natural light from windows, or spillover light from an adjacent lobby or other room. In general, ambient light is non-controlled light. It's not bad, as many churches want ambient light to fill their rooms. However, the more ambient light you have the brighter projector you need.

2. **House lights.** Do you have lights over your audience? If so, can you dim them or turn them off? House lighting often lights up your walls and can make environmental projection a

tough process. If you can turn them off or dim them to only 20%-30% of their intensity, there's a good chance the projected image will pop. Bright house lights can wash out the image and make projection a very difficult task.

3. **Color of walls.** This may seem like a weird factor, but projecting on stark white walls or a flat white curtain will reflect the most amount of light, which will result in a brighter image for environmental projection. If your walls are a more saturated color, it's still possible to get that pop, but you'll need to compensate with a brighter projector.

4. **Size of image.** The bigger your projected image, the brighter projector you will need. You need to achieve a standard brightness per square inch. Therefore, the more inches you are projecting, the more brightness you will need.

5. **Use of fog or haze.** Most often, environmental projection is "front projection" as compared to "rear projection." When front projecting (or when the projector is in front of the projection surface) you need to be aware of the amount of fog or haze in the air, as it will create density in the air and light will be lost before it reaches to the projection surface. If there is any smoke, haze, or fog in the room, it will trap some of the projected light and lose brightness as those lumens are caught in the fog, and lost before it gets to the surface of your walls.

Contrast Ratio

As I mentioned during the comparison of projection technologies, contrast is one of the most important factors in projecting high quality images versus low quality images. Contrast deals with the amount of difference there is between black and white. The higher the contrast ratio, the better the chance of having a good black, which will make whites appear brighter.

Since projection concerns the reflection of light and the light source is always on (your projector lamp), light will always be emitting from the projector. This is projection black. Projection black is the term used for the slight amount of light (or gray) seen when projecting nothing. A higher contrast ratio will allow for less projection black and appear as if the lamp is not on (that's the most ideal situation).

A professional (or projection purist) would argue that contrast ratio is the most important attribute of a projector, but it's one of the highest drivers of price because it is directly related to the quality of lenses, glass, mirrors, LCD chips or light source. Although I agree contrast is a valuable measurement, I would not consider it the most important factor. The main thing you need to know when looking at contrast ratio is that a 10,000:1 contrast ratio is better than a 1,000:1 contrast ratio.

Lens Shift and Keystone

Traditionally, environmental projection requires that projectors be placed in areas that aren't dead center in front of the projection surface. Often this means you're projecting from the ceiling or from an off-angle to the wall, fabric or screen. With this in mind, you need to consider lens shift and keystone.

Lens shift is the ability to "shift" the image up, down, left or right. Some projectors only allow up/down shift and others allow all four directions. The benefit of this feature is greater flexibility in where you can mount the projector body.

Have you ever had an image that looks like a trapezoid instead of a square? That happens when the projector isn't perpendicular to the screen surface. When your projection is off angle, the shape is no longer a square. Keystone correction is the process that turns the trapezoid into the correct aspect ratio and shape of your image.

Most projectors include keystone correction. There is horizontal keystone correction, which fixes a trapezoid that has a greater left or right side, and vertical keystone correction, which fixes a trapezoid that has a greater top or bottom side. Most projectors include a vertical keystone correction, but not horizontal.

Ideally, for environmental projection, you want both horizontal and vertical keystone correction. This allows you to place your projector in a location that isn't perfectly perpendicular to your wall and still project a flat, square image. The greater the keystone angle the more flexibility you will have the during alignment process. For example, a projector that can do a 12-degree keystone correction isn't as good as a projector that boasts a 35-degree keystone correction.

Lens shift is always preferred over keystone correction. Lens shift doesn't distort image quality whereas keystone correction can often result in less clarity or a blurry image.

The Lens

Finally, when looking at what type of projector you need, consider the lens options. Lower cost projectors don't offer interchangeable lenses, which means you are stuck with the factory standard. The projector lens allows you to get the size projection image you need. Every lens will have a ratio to the number one. This factors the ability for a single lens to zoom in or out. If the ratio is not a variable ratio, then it's a fixed lens (example 1.0-2.0:1 is a zoom lens and a 1.2:1 is a fixed lens). As a rule of thumb, most smaller body projectors without interchangeable lenses include a 1.8-2.6:1 lens (but always investigate this before purchasing). Keep this in mind when you are doing your calculations. The wrong lens may mean that you're shooting an image too large and loosing valuable brightness by projecting in unnecessary areas, or you won't fill the wall because the image won't project as wide as you need.

On larger projectors (more expensive), you often have the ability to purchase different lenses with different ratios. Some projectors that have interchangeable options don't even include a lens. I mention this to give you some peace of mind when you ask an AV company to price out projectors and they give you one price for a lens and another for the projector. Interchangeable lenses are very expensive.

Remember when I told you this wasn't a mathematical formula and you were in the market to make gourmet crepes instead of just plain old pancakes? Well, for this one piece, you've got some math to do to find your ratio. That means you need to whip out a calculator and figure out what lens you will need.

Here's the magic formula:

LENS = DISTANCE ÷ PROJECTION WIDTH

Let's break this down. In order to find the exact lens required, you need two measurements: distance (from the front of the lens to the projection surface) and width (the width of your projection surface). In some cases this formula doesn't work exactly. The most common situation in which this formula isn't exact is when a projection surface's height is greater than its width (or any time you need to over project on each side of your wall). I talk about this more in the next chapter, but there is incredible value in hiring a consultant for this install.

There are a number of great iOS and Android apps that make this process simpler. One thing to keep in mind, whether you're doing the math old school or with a lens calculation app, is that the aspect ratio is native to your projector. If you do the lens calculation for a projector that is native to 16:9 aspect ratio and you want to show a 4:3 image, your math needs to be completed using the least common denominator of the 16:9 ratio.

For example, if you want a 4:3 image that is 16-feet wide by 12-feet tall on a projector with a native 16:9 aspect ratio, you will need to do your math based on a 21-foot wide by 12-foot wide image. This is because the projector's internal image-processing chip can't add pixels to the top and bottom of the projected image, so in order to get a 4:3 image out of a 16:9 chip, some of the pixels on either side of the image will not be used.

Overlooking this detail has resulted in numerous miscalculated projection lenses over the years, and I'm hoping you can learn from those mistakes.

In short, remember that every projector is different and the lens you need may vary from room to room. There isn't a one-size-fits-all projector and you'll have to try a few out before you find the perfect one.

CHAPTER 8

INGREDIENT THREE: MULTISCREEN PROCESSOR

At this point in our recipe, you've been able to generate and modify an image, output that image and project an image. However, this is only for one screen. Environmental projection traditionally uses two or three projectors in order to get the ultra wide "immersive" effect.

In most gourmet recipes, there's always a secret ingredient or a special sauce that holds the whole thing together. The multiscreen processor is that ingredient in environmental projection.

There are several different companies that make multiscreen processors and they can range in price from $300 all the way up to $100,000. However, I am going to make this as easy as possible by recommending two products: Matrox's TripleHead2Go (at the time of publication, they cost about $350) and the DataPath X4 (at the time of publication, they cost about $1500).

Multiscreen processors allow you to take a single output from your computer to process images for multiple separate projectors. These processors are basically external graphics adapters that allow you to add monitors on your desktop or laptop computer. The best part of a multiscreen processor is that it allows you to fake your computer into thinking it's one output. This is important because you can leverage the power of software like PowerPoint or ProPresenter without needing an advanced VJing system to output over multiple video cards or graphics processors.

Plugging directly into your computer, the multiscreen processor is an external video splitter, to use the most common of terms. Every professional video tech just yelled at their e-reader or computer screen when I used that phrase, but that's okay (it's them who will look funny yelling at a screen). When you plug this into your computer, you can set your output to be an abnormal video resolution such as double-wide, triple-wide or even larger.

Splitter vs Processor

You may be wondering why I said some people are yelling at their e-reader–it's because I used the term "splitter". It's important to know that there is a difference between a video splitter and a video processor.

In short, a video splitter will simply take an input (say from your computer) and repeat the signal into multiple signals, thus splitting the signal into multiple cables. A splitter replicates the original signal, it does not actually split the signal into different parts.

A processor actually chops up the parts into different and unique outputs. I love brownies, so let's use that for an example. A video processor looks at the entire brownie pan and specifies which areas to put on a plate. If you were to split the brownie pan into three equal sections, the plates you place those sections on would be your video screens (or projectors) and the brownie would be the computer output. Confused? Let me try it this way, too:

Let's say you want to output using three projectors. A common video resolution for your projector may be 1280x720 pixels (a common 720 HD signal). If you were to use a splitter, your computer would send a total output of 1280x720 and each projector would individually see the exact same 1280x720. Instead, if you were to use a multiscreen processor, your computer would send a total output of 3840x720 (this is three times the width of a normal output) and each projector (after passing through the processor) would individually see a different slice, resulting in a 1280x720 projected image. Each of the three projectors would see a different signal, just like it would be impossible for you to put one section of brownies onto three separate plates.

Computers With Multiple Video Outputs

Some of you may have computers that actually allow you to plug multiple additional monitors into your system. This doesn't mean you can get away without a multiscreen processor. In most circumstances, these additional outputs actually result in different video card outputs and the computer sees them as separate outputs. For multiscreen to work with the most common presentation software, you have to trick your computer into thinking it's only one output

(this is what the multiscreen processor is doing, turning one output into multiple displays.)

Most dual-window presentation software don't have the ability to span multiple video outputs, making environmental projection impossible without networking multiple computer systems together. When you span multiple computers, most software requires additional licenses, which inevitably increases the cost beyond the price of a standard multiscreen processor.

I recommend doing a decent amount of research on these products before buying. Some options have a very steep learning curve, while others are fairly straightforward. I would hate for you to spend $20,000 on a solution that could have been achieved for $500. Environmental projection (unless you only need a single projector) can't be done without a multiscreen processor in most rooms.

I have begun to see a few video cards (NVIDIA and AMD specifically) that act similar to a multiscreen processor in recent months. As technology continues to progress and innovate, I expect more to arrive on the market in time. It will be vital for you to research the specific video card on your Mac or PC, to know if it's designed to operate like a multiscreen processor across multiple outputs. You need the presentation software to actually see the multiple displays as one output, rather than multiple outputs.

CHAPTER 9

INGREDIENT FOUR: CONTENT

Content is the most important ingredient of all. To illustrate how important content is, I'm going to share a story from early in my career. I had just started Orange Thread Media and we were hired to provide the production elements for a dinner here in Nashville. I spent almost two weeks planning out all the expenses and details, making sure we had all the right equipment to make the night spectacular, all the while staying on budget so we could pay the right people and make a little bit of money. I had meticulously planned out every detail: what projectors and screens I was going to use, the crew I was going to hire, the specifics on lighting and cables and where I was going to get the truck to deliver all of this equipment.

There was a slight problem with my budget. I had forgotten one small thing: fuel for the truck. Because I didn't plan ahead of time I had to sacrifice my own pay. Environmental projection is in many ways a truck (or at least a vehicle) that needs fuel to run. You can only do so much with a giant blue screen or the template backgrounds from PowerPoint ... but even that is content!

Content is king. You don't need to wait until your media library has tens of thousands of images, but you do need a media library of some sort. Your media (or content) is where the story lives. Content is the environment part of "environmental projection."

We'll talk more specifically in Chapter Twelve about what to look for when finding the right pieces of content, but there are a few things to consider when beginning to build or expanding your media library.

First, make sure you select media with the right aspect ratio (or something close to it). If you're using three projectors (most commonly referred to as a triple-wide setup) and you try to play a piece of media that was designed for only one screen, you are going to have to stretch that content a tremendous amount to fill your projection surface. There are two issues with stretching content with the wrong aspect ratio: you will loose quality and everything will be distorted. Distortion is a big issue if you're trying to show an image that uses circles in particular (stars, particles, etc). These objects become oblong ovals and in most cases you lose the visual impact you were going for. In this situation you would want to find triple-wide content or media. (As a bonus, I've provided a few links to free triple wide media content. You can view all of that at http://LukeMcElroy.com/EPBook-Bonus.)

Second, you want to make sure you have a variety of content, including motion and still images. One benefit of environmental projection is the ability to create dynamic atmospheres that lead to more intimate worship experiences. If you use the same ten graphics week in and week out, the environment is no longer dynamic. In fact, it's very stale and boring. You may as well go back to that plain white wall I talked about in chapter one.

There are a number of incredible artists and creative folks making great content for the environmental projection community. If you've never actually browsed through the content library of TripleWideMedia.com then you're missing out. They have the largest online collection of multiscreen graphics and media and their slogan is "one purchase, three sizes", which means you get three sizes with every purchase. Instead of having to buy your content twice, you can get it from TripleWide Media and immediately download a single-wide (single screen) version, double-wide version and a triple-wide version at no additional cost!

The Mad Scientist

That's it! The four main ingredients for making your own environmental projection concoction. Let this chapter (and entire book) serve as a checklist when you start this process. Hopefully, I've been able to shed some light on the scary, mystical world of projection technologies in a way that makes sense.

As I mentioned, environmental projection is less math and more art. It's an art that doesn't include instructions like "just add water" to make gourmet pancakes. Rather, it requires a chef to perform a very intricate process to produce the most amazingly crisp, yet unburned strawberry crepes. I want you to be known for your amazing crepes instead of flavorless pancakes.

Environmental projection is going to take a lot of fine-tuning before you get it right. Think of a mad scientist in his lab. He never gets it right the first time. There may be some explosions because you mixed two chemicals that don't work well together. Maybe it's the moment you understand that to make the budget work out you have to go with LCD instead of DLP. Or, it may be when you first put up a motion background you're used to only to realize it's way too fast to use for environmental projection.

I love mad scientists. I love how they are obsessed with the learning part of figuring something out. They think outside the box, often times they have understudies to help them out in the process, and every mad scientist I see on TV seems to really love what they are doing.

Become a mad scientist.

Learn. Explore. Create.

CHAPTER 10

THE FIFTH INGREDIENT

Fifteen and BMWs

My dad had the most amazing car when I was growing up. As I think back to it now, it was quite impractical for our family, which is probably why he eventually sold it. It was a BMW Z3. A silver convertible. He kept it inside as much as possible, rarely driving it when it rained. It was in mint condition and a beautiful car inside and out. It was also a very powerful machine.

When I turned 15 I took a test, which the state of Georgia gave me a passing grade on, and I got a driver's permit. I was so excited that I could drive! You probably remember that day. I wanted to be William Wallace from *Braveheart* and walk out of the DMV yelling, "FREEDOM!" because I was one step closer to being able to drive anywhere I wanted, whenever I wanted.

On the way back to the house I couldn't wait to ask my dad for the keys to the new BMW ... I mean as far as I was concerned, I was an expert. But I wasn't. In fact, I had no clue how to drive a car. I just knew all the driving laws and best practices.

I had to start slow. In fact, I had to drive this beater of a car; a green early 90s Volvo sedan. Okay, it wasn't actually a beater, in fact it was a fairly decent car, but in comparison to the immaculate, brand new BMW Z3, it sure seemed like a beater. Though I didn't agree with them at the time, my parents would have been crazy to let me learn how to drive on the BMW. Instead we started slow on the Volvo and eventually I was able to drive "The Z" (as we affectionately called it). That was one amazing car and I'm not even a car guy!

The Fifth Ingredient

Environmental projection is like the BMW Z3, and you are the 15-year-old with a learners permit expecting mom and dad to give you the keys. You need to learn how to drive, and how to drive well, before you can take the car for a spin around the block.

The fifth ingredient in the environmental projection recipe is training and education. It's the journey you must go on to fine-tune your recipe and measure out each ingredient. I'm pretty big on hiring an expert to teach, train, consult or lead me when I'm learning to drive. I've recently been sharing at seminars and conferences that 10 percent of your visual worship ability is your knowledge of tools and technology and 90 percent is your process and passion for what you do.

In his New York Times bestselling book, *Start*, author Jon Acuff writes, "Never compare your beginning to someone else's middle"[1].

That's why the fifth ingredient is by far the most important piece of the environmental projection process. You can't just hang a few projectors and throw some media up on the walls. You have to begin a process of seeking, learning and wrestling with what speaks to your own communities and congregations. Creativity absent of context can lead down a dangerous path of entertainment-minded worship services.

Here's something I've learned on my own journey of environmental projection and worship. **Unforgettable experiences don't just happen.** They take planning and intentionality to execute. Simply having environmental projection in a worship service doesn't mean it's going to create a life-changing atmosphere. Taking that last statement one step further; pouring the four other ingredients into a bowl doesn't make environmental projection.

Understanding the "what" of any system is important, but you need to wrestle with the "why" before you can really engage your heart in the project. I don't want you to become button pushers, or spacebar monkeys. I want you to be excited about becoming mad scientists. And in order to become a mad scientist, you have to learn from one.

This doesn't mean you have to hire someone to assist you in installing environmental projection.

But I highly recommend it.

This doesn't mean you have to go to a conference to learn the art of a great VJ.

But I highly recommend it.

It does mean that there is incredible value in seeking advice from the experts and hanging out with them as much as you can. Maybe they will rub off on you and your visual worship journey will be ignited and the why behind environmental projection will be revealed.

Seek and You Shall Find

When it comes to environmental projection training and resources there are a slew of places you can go to learn. I know you have a yearning for learning or you would have skipped right over this section. I'm not saying environmental projection is rocket science, but it is a very dangerous thing to implement into a worship service without intention and training. You're dealing with an environment where people meet God. That's not an environment you want to jump into too quickly. There's an incredible responsibility that comes along with it.

My overarching encouragement is to seek out great resources. Here are a few ways you can learn from an expert:

1. Attend a conference or seminar.

There are so many great conferences that touch on visual worship. Just Google "Visual Worship Conference" and you're sure to get a whole list of them. What I love about going to conferences is the amount of concentrated information you can get in a few days of classes and workshops. It's like drinking water from a fire hose. Bring a notebook and be ready to fill it up! Ask tons of questions when you can and seek to meet other people in your field. Conferences are also a great way to be exposed to new technologies and/or companies that are making the process easier. Who knows, you and I may even get a chance to meet face to face at a conference! In 2013, God placed in me a call to start a conference for those who wanted to see creative arts in the Church as a vehicle for God to transform lives. Environmental projection, of course, plays into this heartbeat. If you're interested, I encourage you to check it out at SALTNashville.com.

2. Hire a consultant or expert.

Inviting in a consultant or expert is like having a customized conference for your team. It gives you the chance to review your processes and plans in-depth to make sure you're on the right track. Though it may seem costly, (when you factor in hotel, car, flights, and consultant fees), the return on investment will be incredible. There's no question, learning from an expert is the best way to learn about a subject ... especially environmental projection.

There's something else I want to mention about hiring an outside consultant. Be careful whom you hire. There are a lot of people who claim to be "experts" but don't have any way to back it up. In particular, if you're going to install environmental projection equipment, I recommend hiring someone who has installed before and can show you several examples of their work. Hiring someone without a fine-tuned recipe is no help.

3. Books and blogs.

First, you've picked this book up ... that's an incredible start. My hope is that you've already learned a ton about this subject. A book is a great way to get a jumpstart on a topic without overwhelming yourself. Though there aren't a ton of books on environmental projection or visual worship available today, I believe that will change. Books are the most convenient of the various ways to learn, and there are two I'd encourage you to check out if you're interested in reading more on this topic. Start here; *The Wide Guide: Blueprint for the Multiscreen Movement.* (As a bonus to purchasing this book, you can download a free copy of *The Wide Guide* at http://LukeMcElroy.com/EPBook-Bonus) It's honestly a predecessor to the book you're reading now, but I go through the eight most common multiscreen setups, sharing diagrams, tips and ways to save money on each. Second, take a look at a blog our team put together highlighting a few books every creative person in ministry should read at http://LukeMcElroy.com/EPBook-Bonus.

4. Ask a colleague.

If you can find someone (maybe another church) who has gone through the process of installing environmental projection equipment before, you can gain a wealth of information. It's nice to have someone who's not trying to sell you a product or charge you for their time to give you some advice. These people may be tough to find, and it may take buying them a lunch to get enough information to make it worth your while. You can also check out some online communities to seek answers. For example, Church Technical Leaders. This is an incredible community of church technical directors, creative pastors, visual worshipers and people of the like that work in a similar setting as yours every week. You may be amazed at the value colleagues can bring to a process like this. They're out to help you, as well! (On the http://LukeMcElroy.com/EPBook-Bonus page, I've listed a few recommended Facebook groups and pages that may be a good place to find community.)

A Cook or a Chef?

I got a chance in 2013 to visit the set of *Hell's Kitchen,* the Fox reality television show hosted by Gordon Ramsay. It really is a great show, and I've become a big fan of the star of Ramsay himself. He's an interesting man who expects nothing but the best from those in his kitchen(s). It's why he yells, screams, returns food that has the slightest issue, and kicks people out of the kitchen. He demands a lot out of people who want to be chefs.

It's part of his process. And that's the difference between a cook and a chef. Ramsay believes everyone can (and should) cook, but it takes training, passion and knowing what you're talking about to become a true chef.

I don't want to just be a cook of environmental projection. Chefs lead, cooks follow. Everyone starts as a cook, but the chefs are those who are persistent and ask questions. They're lifetime learners, they're leaders, and they're constantly trying to get better.

That's what environmental projection's fifth ingredient is all about, learning to learn. Eventually, you're all going to be chefs driving silver BMW Z3's with the top down ... but it's going to be a journey that takes some time.

CHAPTER 11

STEP BY STEP

Disclaimer: One Size Does Not Fit All

Now that you're all passionate about becoming chefs who drive BMW Z3's we can actually dive into the process of setting up environmental projection. As I mentioned in great detail in the previous chapter, this isn't a mathematical formula, but rather an artistic process that requires refinement.

No two rooms are the same, which means there isn't a one-size-fits-all environmental projection setup. So feel free to make modifications to this setup to fit your room. Some rooms have low ceilings and wide walls while others will be tall and narrow. I've seen rooms with fully controlled lighting to rooms with too many windows to count. That being said, it's important to remember that this process isn't simply a cut and paste setup for everyone.

Step 1: Evaluation

The first step in setting up environmental projection is to evaluate the room. This may include a survey of the room's architecture, the use of lighting, the shape of the room, the angle of walls and many other factors. What you want to do here is get your head around a solution that makes sense for the community that worships in this room. An important step in making environmental projection effective is creating an environment that speaks to the specific community. A good consultant will ask for pictures, bombard you with a ton of questions about your community, or even experience a worship service before making any decisions.

Don't skip this step. Those who do spend way too much money on gear they don't need or install projectors that don't work for their room. During this phase you are going to want to evaluate the amount of ambient light you have to compete with, as well as determine where you want to project. These decisions will help you identify the brightness of projector you should consider and where you need to project from.

Step 2: Measurements

Next, you need to start getting measurements of the room. Remember the formula I gave you for calculating a lens in chapter seven? Those measurements are going to help you fill in the formula. Since you determined brightness in the previous step, this step is more about how many projectors and what lenses are needed in order to make environmental projection a reality. You may want to pick up a laser measurement tool to save time if your measurements are going to be over 25 feet (On the http://LukeMcElroy.com/EPBook-Bonus page, I've listed a few measurement tools you may want to consider purchasing.). Here are a few measurements you will want to get:

- Width of entire projection surface.
- Height of projection surface.
- Distance from back of the room to the projection surface.
- Width of individual areas for projection.

Go back to that formula from chapter seven and you'll be ready to turn these measurements into something helpful: identifying the correct lens. If you're having trouble deciding where to place the different projectors, don't worry, the next step should help you narrow down your choices.

Step 3: Natural Intersections

When placing projectors, you want to take into account the design of the room and look for natural intersections, or places in the room (or on the walls) where you can hide the alignment of two projected images side by side. Clear as mud? Let me try it this way. When you place two projectors side by side, it is virtually impossible to perfectly align them in a way that you don't know where one image stops and the other begins. Therefore, if you can find an angle in the wall, a column, a vertical frame or something else that breaks up the image, you're in luck. The goal is to hide the edges of the projected images and make them appear to be seamless paint on the walls.

If your room doesn't have any natural intersections (if you have a 270-foot wide solid white wall with nothing else on it, for instance) you will want to learn more about a technology called edge blending. I won't go into it here, but edge blending will allow you to blend the edges of a projected image in order to remove the need for a natural intersection.

In short, this step allows you to use natural breaks in the room to hide the transition from one projected image to another.

The end result will be one uniform image that spans the entire width of your wall.

Step 4: Buying Equipment

Now that you know your lenses, and you know your natural intersections, you're ready to buy equipment. This is a good time to go back to chapter five and review the four ingredients. I've made it convenient for you by listing them here:

The Four Ingredients:

- Software
- Projector
- Multiscreen processor
- Content

Now, the daunting task of finding a good place to purchase all of this. I highly recommend going through someone local (that you can call up at any time and have them come over) or someone who has a great reputation (and they will do whatever it takes to help because they can't risk tarnishing that reputation). This could be a local AV (Audio/Video) integrator or a reputable national vendor. Regardless, get two to three quotes and tell each bidder you're doing so. This will teach you a few important things.

First, these AV dealers will be forced to give you a great deal and not just a good deal because they know you're not going to just give them the business. They're going to have to be the best solution and most often price is a huge factor.

Second, price isn't the only factor. In fact, you may choose someone who is actually more expensive than others, but this process will motivate them to sell you on their expertise and knowledge. This is critical. You want someone who doesn't just sell gear, but will be a great resource for troubleshooting, knows everything there is to know about video technology, projectors, etc. and has passion and heart for creating powerful visual worship environments.

Finally, telling someone you're considering another company will move them to work harder and give you a longer guarantee or make you aware of other warranties available. All of these benefit you and help you make informed decisions.

He With The Most Information Wins ... Every Time.

Every AV integrator is different, and even though I've had the privilege of getting to know many over the years, (several are very close friends of mine) you need to find the one that is best suited for you and your community. Here are a few things to consider as you evaluate AV integrators.

First, let them show you that they're faithful in the little things. This could be as simple as purchasing a few small things from them to see if they live up to their customer support promise. If they're not faithful in the little things, you can't guarantee they will be after you make a large one-time purchase.

Second, ask them if they've offered you their best price. I got this little piece of advice reading Dave Ramsey's *Total Money Makeover*[1], and it's become a very effective strategy for me over the years. Just because it's the price they give you, doesn't mean they can't get you a better one. The worst they can say is "no", but you'll be surprised how many people may give you a discount just because you ask for one. I can't guarantee it will work, but I can guarantee this, if you don't ask, you won't get one.

Third, get references. Even if you don't know anyone that has worked with this company, ask the sales rep or your contact person for the names of a few churches they've worked with in the past. Spend 15 minutes looking up a phone number or email and ask that previous customer about their experience. The best way to know the future success of any person or company is to follow the breadcrumbs of success they've left behind them. Following those breadcrumbs maximizes your opportunity to choose the best AV integrator.

Finally, avoid buying from a mass market, big box store. I'm talking about Target, Wal-Mart, Sam's and others. They don't know how to support your product and you will never receive any kind of expert advice. Good AV integrators will accept product returns if they feel they sold or recommended the wrong product to you. If an easy return policy is the reason you want to buy from some online mega-store, ask your AV integrator how they handle equipment that doesn't work.

Take your time with this process. It won't be done overnight. I have walked with churches that took several weeks to a few months to lock down all the details. You're dealing with very meticulous processes that can be exhausting. (Remember the fifth ingredient).

Step 5: Install

Now you're ready to install your equipment. Of course you may have someone else doing this for you, but regardless, it needs to be done. There's not much to say here, except take time to map out a game plan before you start drilling. Many installers call this a master plan. This will include a whole lot of details.

I'm not an installation expert, but I hate it when you have to move projectors that have already been installed. You know what they say: Measure twice, cut once.

Whenever you are hanging expensive and heavy equipment over the heads of your congregation, you really should consider using a certified and insured installation professional. Even though I have friends who have done live event video production for decades, it's still an industry standard to hire certified people to come in and attach your hanging points to the ceiling. Safety is vital in this process, and I highly recommend doing your homework as you prepare for the installation process.

Make sure you have tested the hanging and rigging mechanisms to support seven times their weight—an industry standard metric.

If you need to run new power or electrical as a part of your installation, double-check that all of the components are wired correctly and that outlets are well grounded. Projection equipment needs ground in order to have a clean signal, otherwise, you will see a constant pulse in your image and quality will be sacrificed. It's also smart to bring in a licensed electrician as you complete your install. You don't want to let power get in the way of a great environmental projection setup.

This is not a process I recommend doing without a well thought-out plan and a detailed look at safety precautions. Even a temporary setup can have its issues. Take your time, develop a plan, get expert advice and always make sure the priority is keeping your volunteers, staff, congregation, property and equipment safe from negligence and impulsive decisions during installation.

Step 6: Alignment

Once you have installed all of the equipment, mounted your projectors and have run all the power/signal cables, it's time to align them with one another. You'll know at this step if the projector lenses you selected were good choices. You will need a few good test patterns in this step. (I've put a few links to Alignment Grids on the http://LukeMcElroy.com/EPBook-Bonus page for you.) I have yet to see an alignment pattern of quality built into the software.

There are three primary test patterns I recommend. First is the alignment grid, a series of lines and high contrast crisscrossing grid-like pattern to make the process of aligning multiple projectors together significantly easier to see the alignment. The goal in any environmental projection setup is to make the projected images appear as one consistent image, and there's no better tool than the alignment grid to make sure that tilt, keystone, zoom, focus and position are all in order. Sometimes your software may have an internal alignment grid, but I've found that the one we provide gives a cleaner image. And since it's actually a piece of "media" instead of internally generated, it will give you a more accurate representation of how media output from your software will work.

The second pattern I recommend is the horizontal color bars. Usually you would want to use color bars when aligning a projector so you can fine tune the color and make sure you're not missing any specific hues. Color bars will not only check the signal cable to see what you're missing, but also the projector to make sure it's processing the colors correctly. When you're using environmental projection, you have to take the concept of color accuracy between multiple projectors one step further; do the colors match one another? I've found over the years that this is impossible with regular color bars. The horizontal bars make it really easy to see the various colors of two separate projectors side by side with little gap between them, making it extremely easy to color correct.

Third, you will want to use a gray bars test pattern. Once you've made sure your projectors are aligned properly and your colors are consistent, this pattern will help you evaluate the evenness of brightness from one projector to another. As I mention in this book, brightness has a lot of factors that cause variable outputs. By using the gray bars pattern and putting various shades of gray side by side, you'll be able to correct and dial in the brightness settings of a specific projector to make sure you have a clean, consistent and smooth image across your entire room.

All three of the test patterns I mention above are must-haves in any projectionist's toolbox.

The alignment step is pretty simple. You're trying to align all the color-coded lines and angles. You also need to make sure the projectors are all set to front projection (this can cause issues later if you purchase content with text on it). As well, you need to square off all the lines and use lens-shift/keystone as needed to make the projected image fit perfectly on the wall.

Warped images appear when your projectors are not keystoned correctly. Also, check to see if your lines are level. If one side is higher than another it will be disorienting once content is on the walls.

If for some reason you purchased projectors that don't have good internal keystone correction, outboard keystone correction boxes are available. Kramer (a respected name in the video industry) makes a spectacular product that can rotate, deal with curved walls and provide advanced keystone correction.

Step 7: Image Masking

Once you get all of your projectors aligned, there's a good chance your image will project onto areas you don't want it to, such as singers, set pieces, the ceiling, etc. This is where the concept of masking comes into play. This process is like making cookies in the shape of a star. You need a "cookie cutter" to remove the parts of the projection you don't want. Instead of calling it a cookie cutter, though, call it a mask.

A mask simply projects black where you don't want an image so that the projection image seems to magically fit perfectly into place with ridges, ceiling lines, balconies, speaker clusters, etc.

Every room is different, which means there isn't a standard mask, and that means you can't just purchase a mask somewhere. I wish it were that easy! Here are some suggestions about masking for environmental projection:

1. **Start with white.** I find masking is easiest if you start with an all white screen. This makes it simple to find the places you don't want light to shine onto. Then you fill in those places with black. Once you have a template, you can either apply it to your entire content library or use it within your presentation software to save time.

2. **Remember the choir.** If you've ever stared into the lens of a projector when it's emitting an image, it's something you'll never do again. A projector can be really bright, so don't make your worship leaders or choir members look straight into the projection image either. Mask them out of your image. It will make the environmental projection look better and they won't hate you every week. (If you have a choir that's too large or projectors that are too bright, you may not be able to shop at the same grocery store you always do).

3. **Mask out screens.** One thing to consider is creating a second mask for the times during a service when announcements,

PowerPoint slides or videos play on your main screens. You don't want competing images during these moments and masking out your screens will make it look like the projection is paint on your walls. Remember that rule of thumb from chapter three?

4. **Feather the pews.** Be considerate of your congregation. If your congregation is in the path of the projector, simply blur out the bottom part of your projection where it hits the floor/seats. This will make for a natural soft edge and allow the eye do the rest of the work down there.

5. **The ceiling matters.** Environmental projection is supposed to look like digital wallpaper, and wallpaper shouldn't bleed onto the ceiling. Ceilings that aren't masked will give your images random cut off points, making your entire presentation appear amateur. Make sure to spend the time to get a detailed mask along the ceiling.

There is a software that is fully dedicated to making masking easier. Designed by our friends at CrowdControlGames.com, is MaskeR-Aid[3]. This software allows you to draw a mask in real time while sitting in your room. After you create the mask, you can go point by point and adjust until it's perfect. Finally, export your mask as a transparent PNG image and import it into your software.

Step 8: Content

You're almost done! Finally, it's time to pick out the right content for your walls. This is the step you will complete every week. You're telling a story with your walls and visuals. So what imagery do you need to tell that story? Remember, this is the fuel of your vehicle. Don't forget to account for the time you'll need to curate great content.

This is where TripleWide Media enters the picture. They are constantly adding new media to their online library and without question they have one of the largest content libraries online.

Aside from purchasing content you can also make your own content. This is fun because it brings the community creativity concept into play. It may be as simple as creating an image for a sermon series that is used during announcements or as detailed as using Adobe After Effects to crank out some custom motion backgrounds.

CHAPTER 12

CHOOSING THE RIGHT MEDIA

Blaming the Medium

I can remember it like it was yesterday. I was shouting a familiar chorus with 5,500 other people in the Atlanta Civic Center as a recognizable worship leader led us in a song many of you would know. (I'm keeping the details hidden for the sake of the worship leader.) He sang with a massive, 50-by-25-foot video screen behind him. It was one of the first applications of multiscreen in a worship environment I ever experienced.

The video was fast and moving at an incredible speed. It was like an oversized marquee displaying lyrics on top of an out-of-control rock concert strobe light, but the moment didn't call for it. It was a worship song. One you and I have probably sung in our very own churches.

The setup was impressive, yet, I have never since been as distracted in worship as I was in that moment with that professional.

I left blaming the massive video setup, thinking that some screens and projectors ruined the night. Why would they ruin such a sacred moment with all this video background junk? It was too much, too noisy, and way too secular.

Then it hit me. It wasn't the projector's fault. It wasn't the screen or the cables that fed the projectors' fault. In fact, it wasn't even the computers or software that distracted me. It was the content. We've all blamed the technology for being distracting, but in this case it was doing exactly what it was designed to do, project media.

Content is where the story lies, which means it's where the impact or distraction lies. I've been part of hundreds of worship environments as a part of live events with Orange Thread Media. I've seen thousands (literally) of pieces of content through my curation role at TripleWide Media and I've even made a few motion backgrounds myself. And I'm here to say the distraction some people see environmental projection as being is the fault of the content being used, not the environmental projection itself! Stop blaming the technology!

We Need Rules

As I look back on my experiences with multiscreen and environmental projection, I can very quickly identify five rules I have made for creating powerful visual worship moments. And, looking through that list, I realize I have broken every rule in some way shape or form. That's probably why I came up with the rules in the first place. Every time I broke a rule, it was a distraction for those who were in the worship service.

Don't get me wrong, these aren't the only rules involved in playing back media, but they do lay a foundation on which we can create a powerful instinct for what looks good in immersive environments. My hope is that these rules become the filter that you sift media through before you purchase it. They've helped me choose the right media for the right circumstance before. Here are my five rules for multiscreen content.

First Rule

For most of us, we find new media when scouring the web for the best deal. We visit countless websites and keep mental notes of our favorite producers. Watching video after video after video, we finally make a decision about which motion backgrounds will be a good fit for a specific moment in our worship service. If you don't go through this process, I highly recommend you start. But back to the first rule.

Let's say you find a motion background you like and you want to use in your environmental projection on Sunday. You log in to your iPad and notice that the motion has a few particles and a solid color. It's perfect. As you preview it, you notice that the particle seems to just lightly drift from the left side of the screen across to the right side. It's peaceful and flowing like a dandelion soaring in the cool summer breeze. So you download it.

You then load it into your presentation software, boot up the projectors and import the newly purchased motion background. Something crazy begins to happen. The same particle that was once an innocent little dandelion is now a meteorite crashing into earth at an unstoppable speed. It's almost got flames coming out of the back of it. Why? Because media appears to move faster on a 150-foot-wide video wall than on a handheld touch screen! Bringing us to rule number one:

RULE #1: Your media will appear to play faster on a larger surface.

If you're using multiple screens there's one key factor to consider when projecting motion backgrounds: speed. In the Atlanta Civic Center. the enlarged screen made the content appear too fast. When our projection medium is enlarged, our media is enlarged. Eventually I learned that you can always speed up a video clip, but it's very difficult to slow one down.

Most of today's presentation software options have some sort of speed feature, but slowing down a video clip or motion background can cause unwanted stutter. Video is made up of a bunch of still images playing back at 30 frames per second (that's what 30fps or 60fps stand for). When we slow a video down, eventually our eye will notice the changing of still images and it won't appear as a movie anymore, but more like a fast slideshow ... causing the motion to skip, stutter or "hiccup".

Keep this rule in mind as you grow a media library and especially when you look at media online. More important than all, make sure to test out your video clips before the worship service starts. You'd rather have a motion background that you paid for and don't use than a distraction in your worship service. I never use media at an event that I haven't tested previously.

Second Rule

Motions can be dangerous territory to journey into, but can also create some powerful moments when used correctly. However, I want to break the myth that you HAVE to use motions in order to be effective. You do not. In fact, I recommend starting with still images. Always.

What?!?! The guy who just said he has made his own motion backgrounds says to stick with still images? I sure did, and here's the logical reason:

Rule #2: You don't always need motions.

When I talked about multiscreen content in chapter three, I mentioned there are two main categories of presentation software: single window and dual window. The major difference between these is the ability to do motions (for the most part). That's why I said a single window software may be a great starting point for environmental projection. No doubt, the dual window softwares (ProPresenter, MediaShout, EasyWorship, Proclaim etc.) have a slew of other powerful features under the hood, but not having one of those doesn't mean you can't make environments full of impact.

When it comes to creating an immersive atmosphere, the right still image or texture can create a powerful environment and may be all you need to tell the desired, deep and rich story. Consider using a texture or a still that has some ambiance in it. Maybe it's abstract, maybe it's literal. By adding texture to the walls, you deepen the opportunity to feel without the distraction of moving elements. A texture will also add a good deal of contrast that often sets the mood for a powerful moment, rather than an otherwise white or bland wall.

The Walk Youth Retreat. © 2011 Luke McElroy.

Take this picture, for example. Instead of simply singing an ancient hymn together, we were able to create a sacred moment by using a still image (and a beautiful texture) to usher people into a virtual cathedral, amplifying the meaning and feeling of the song.

The other benefit of still images is cost. On any of the many websites that sell media, still images are a fraction of the cost of motion backgrounds. Many times, producers will make their media in both a still image and a motion background. Start with still, move to motions later. I've been a part of entire worship conferences that just use still images for environmental projection and they're often more powerful than those using motion backgrounds.

Also, I find that still images allow you to bring the narrative into the visual environment without so much distraction. Sometimes, when you're projecting large and incorporating moving subjects, it begins to look like an IMAX movie instead of a meaningful, deep story. However, with a still image you can introduce a concept, create context and let the worshipers or congregation live their own story in the setting that has been established. This is most commonly seen with cathedrals, or natural photography of landscapes and mountains. It may even be a brick wall or a perspective of Times Square in New York City.

You don't always need motion. Just like you don't need drums in every song. Visuals are much like music in that you need to be dynamic for impact. When you're looking for media to add to your visual library, I encourage you to seek out still images, textures, contexts, non-moving narratives and powerful visual landscapes that don't have movement to them. These will add depth to your presentations and lower the risk of distraction.

Third Rule

I may have just added more fuel to the fire in the great debate of motions versus stills, but there's still a third characteristic about media you need to keep an eye on. Whether motion or still, every visual you use will tell a story. Even a solid yellow still image can tell a story. Here's how I know:

Rule #3: Emotion and energy are elaborated upon with color.

Psychologists have been studying color for a long time. Why is it that each of us has a favorite color? How come this football team has a red jersey and the other team has a black jersey? Going back to the story of Moses on Mount Sinai, why was color important? In 1953 there was a major breakthrough in the study of color. A psychologist by the name of Charles E. Osgood had a theory that color affects our psychological makeup and emotional reactions to certain situations[2].

After years of testing and thorough analysis, including hundreds of individual and group studies around the world, Osgood wrote this statement in his final conclusions: "Specific assets of certain stimuli [including color] elicit distinct, innate and unconditioned responses."[2]

Did you catch that? What he is saying is that color (stimuli) causes an uncontrollable response or reaction! This isn't just something for psychologists, but a huge breakthrough for anyone creating, controlling or curating a visual environment. What this means is that the colors we use in a room will indeed affect the way in which we respond in the given moment.

This isn't just a rule for multiscreen visuals, either. It's a principle that all visual leaders, lighting directors and artists must adopt: Color affects our emotions.

The same is true in music. You've probably been to a symphony before, if not I'm sure you've heard a classical tune here or there. A friend of mine was given two tickets to see the Nashville Symphony a few years back and invited me to join him. It was an exhilarating experience. The music took twists and turns, leading me through a myriad of emotions and feelings. Now I'm no music expert, but I did go to college at a school that had a pretty renowned music program, of which, many of my close friends were a part. Music has an ability to affect our emotion. Specific chords can elicit emotional responses, and great songwriters use these chords at strategic times to tell very specific stories and engage our hearts in the story.

I think this happens a lot with our very own worship music. Songs touch our lives because they are designed to lead us somewhere and engage us emotionally. Visuals work much the same way. However, we often allow the story and emotion from the visuals to tell a different story from that of the music. Remember, every visual is telling a story because every visual either has color, or is strategically lacking it (which still tells a story).

Take a look at this chart and you'll see some of the basic emotions color helps us feel. When used in the right moments, color can amplify the story of the song or the meaning of the moment. You're elaborating on an authentic feeling that helps your community engage in deeper and more authentic worship experiences.

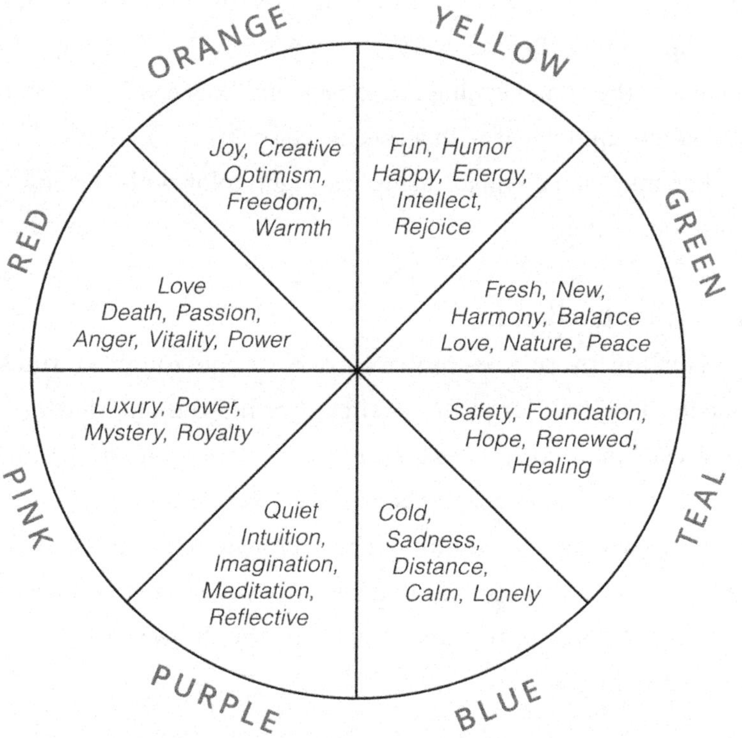

Color & Emotions Chart. © 2011 TripleWide Media.

Since color has an ability to enhance the story we tell, then how is the color of the story you're trying to tell on Sunday morning enhancing that story? Do you have a library that covers the whole spectrum or are you leveraging one specific color pretty heavily? Consider using only solid colors or monochromatic textures one Sunday and see what happens. If you have a lighting team or someone who can control color within your lighting system, make sure there is a consistent theme across the color palettes.

In these situations, when lighting is present, it's critical to remember that lights are telling a story with color, as well. If the colors aren't the same, or complementary, then you're telling conflicting stories! Before every worship song, I ask the question, "What is the song begging me to feel, and how can I support (and ultimately amplify) it with a specific color?"

For musicians, equate using the right color with turning your guitar amp up to 11.

Fourth Rule

The fourth rule is probably one of the toughest rules to master. I still fail at it all the time, but it's the one that has the greatest kingdom impact long-term. It's not something superficial that you can check at the surface level. You're going to have to dig down deep into your spiritual knowledge, your theological background and your familiarity with your visual library. The fourth rule begins to move you from the "how?" to the much deeper "why?".

Rule #4: Make sure the message is stickier than the media.

The goal of environmental projection or visual worship isn't to entertain. You're not looking for "wow" moments for the sake of creating wow. You create "wow" moments so people have the opportunity to be connected to their Heavenly Father in a more intimate and powerful way. This means media can become a distraction during that wow if you're not careful.

Visual worship must move people to think differently and walk out changed from how they walked in. Otherwise, it was just a bunch of cool media on the wall that had no spiritual impact. You need to evaluate every piece of media to make sure the message that's being shared, or the story that's being presented, is much stickier (and far more memorable) than the media that was used.

The best artists in the world are able to tell a story through their paint, pixels or even a pen. The best visual worship artists seem to intersect the pixels of their medium with the theology and grand narrative of the moment to create life-altering, contagious visual environments. I'm talking about the artist who is able to take an image of the cosmos, merge it with a smoke-like texture, and communicate something that only God could speak. This comes from understanding that your media can't be more "tweetable" than your message.

A few years ago at Easter, my home church created quite a visual environment. Joe Blair, production and creative director for Fellowship Bible Church just south of Nashville, TN called me up one day and said, "I have a crazy idea and want to see if you can help!" He wanted to do 180-degree environmental projection AND project on thc cciling.

Before I continue this story you need to know that the 10-year-old kid inside me had wanted to project onto the ceiling for a LONG

Fellowship Bible Church. © 2012 Luke McElroy.

time and I felt my time had finally come! So we grabbed lunch and he shared the whole vision with me.

Over the next few weeks, we planned where we were going to project from and what we were going to use as screens. I learned that if you're going to suspend screens on the ceiling, they have to be flame retardant and water has to be able to pass through them easily (in the event of a fire of course).

When it was all said and done, it looked incredible. Here's a picture I snapped from my phone that morning when people were walking in.

However, we didn't do it to be able to say we projected on the ceiling when the service was over. We didn't do it to post on Facebook and Twitter expecting people to like, share and retweet. In fact, among the Facebook comments we received, not a single one was about the media. They were all about the message. You see, our pastor was in the book of Luke, talking about the parable of the mustard seed.

He made the connection that the mustard seed is like the Kingdom of God. It started small and through time it has gained momentum. After Jesus died, it hit critical mass and became an inescapable force in the world. He said that as time goes on, the Kingdom of God will be like an unstoppable tree growing and growing and growing. It will eventually reach the ends of the earth and we will all worship under this great tree one day.

Do you know what we did? We projected a tree. Growing. For 45 minutes while my pastor spoke. It eventually (after about 18 minutes) started to show up on the screens in the ceiling and continued to grow more and more. By the end of his message the tree nearly covered the entire canvas (both the environmental projection and the screens in the sky) and we worshiped together under the great big tree.

Fellowship Bible Church. © 2012 Luke McElroy.

That's an example of making sure the media comes second to the message. That weekend we gave our body a picture to hang their hats on for a lifetime. We gave them an experience that will help deepen their engagement with the local church, their interaction with the Holy Spirit, and a greater understanding of the glory and kingdom of our great God!

What story are you wanting to share? How does the media in your library lend itself to be a sticky and memorable message? Do you have the right media to tell that story? If our ultimate prayer is that our media and art will impact people's hearts, then we need to tie our media to a message or a moment. People will remember a message, but they often won't remember a piece of media.

Our responsibility as artists of the visual environments for our churches is to ensure the media we use supports and edifies the message being shared, rather than pulling attention away from it. In the story of the great big tree, we didn't add a running groundhog. That would have been distracting. We just had a tree.

Let's move from people who push buttons to artists and ushers of powerful moments that engage our communities in the depth of

God's love and never ending grace. Let's be curators of environments that lead people to a greater understanding of the life-changing power of Jesus Christ. And let's make sure that the message we're telling the world (with both our lives and our media) is stickier than the media we're using.

Fifth Rule

Lastly, you need to make sure you don't overdo this whole visual worship thing. I've been relating visuals to music throughout this section and it's because of the incredible correlation between the two. Just like in music, visuals need to be dynamic. However, as technicians, you sometimes get confused as to what that means.

Let me clarify: **Turn. It. Off.**

Rule #5: Less is more.

Early in my visual worship journey I assumed bigger was always better. I have learned over the years that this is far from the truth. Don't get me wrong, a large video surface can be very powerful and isn't inherently bad. The point I'm trying to make is that some of the most powerful and intimate moments I've had with God come right after the big moment, when everything gets really small and really personal. I'm talking about those moments when everything disappears and is turned off for a song. A visual breath.

This is the vulnerable space in time when we don't have visuals to communicate for us and we're left to create the visuals for ourselves in our own imagination, thus creating an experience that's between us and God. That's where some of the most beautiful moments come for me. I have to ask myself all the time if I'm doing everything in my power to allow those moments for other people.

Every once in a while, I need to turn everything off and allow for space and stillness in the room.

I just got back from a great trip to Seattle where I was helping a church on the peninsula create a visual environment for their Easter services. While I was there, I encountered a breathtaking view of Mt. Ranier on my drive home from the church. One day, the entire team and I were headed to lunch, and I asked the intern, "What's it like living in a place where that mountain screams the glory of God every single day of your life?" His response amazed me. He said, "I don't really notice it all that often."

The issue was that he was so close to it and saw it so regularly that he never stopped to think much about it. Later in the conversation, he did admit that if he were to ever leave Seattle, he would likely miss the sight, but that would have only been after being away from it.

The same is true with our visual worship environments. If we're using environmental projection every single week and in every single worship service, it's eventually going to become so regular that no one notices the mountain in the room. This means no one will get the impact out of it that it's designed to deliver. But, we want environmental projection to be full of impact. You wouldn't have read this far if you didn't want that.

This really rang true for me a few years back when I was blessed to go to one of the most beautiful places in the entire world, New Zealand. Our family was traveling around South Island exploring some of the beautiful fjords and sounds of the area. It was breathtaking. I wish I could describe the beauty. On one trip, we were invited to get out in a kayak and paddle around the still blue waters.

All of the sudden our guide told us to stop. He asked everyone to put their paddle across their lap and not move a muscle. Then he instructed us to sit in silence for 60 seconds and just let the sounds of the area settle in. It was the most beautiful thing I had ever heard. We had been in New Zealand at the time for about six days and our trip was coming to an end. I remember writing later, "It's amazing how beautiful life gets when I shut up." (Many of my friends may say the same thing today if you ask them!)

This is so true in terms of visual beauty; life becomes more beautiful when we turn it all off sometimes. It's that moment, when we invite people to ponder, to imagine and visualize. It's that tender moment in a song when a musician holds the note just a little bit longer than you're expecting. Life can't be beautiful when we're making noise all the time, even if it's visual noise. Sometimes it's in the visual silence that we notice the beauty.

I just want my story to be one that beckons people to begin a conversation with God. I want my life and my visuals to reflect that in some way, which means I'm really aware of the tender moments in my visual worship leading. Let's embrace the art of simplicity and the art of subtleness. Let's be okay with turning off environmental projection for a song, a week or even an entire season. Let's be okay with letting stillness speak on its own and letting our environments tell great, powerful and dynamic stories, even if visuals aren't present.

CHAPTER 13

PARTING THOUGHTS

It's Not Going Anywhere

That's right. It's not going anywhere. Environmental projection isn't just a fad or the new, hip, cool thing that is luring in the next generation. I argue that it's here to stay. We've seen how stage layout and worship spaces have evolved over the years. I'm not saying that our worship space won't morph again over the next several decades, but I am saying that visuals that help create immersive environments are here to stay. It may not be "environmental projection" and it may not be multiscreen, but I do believe that the global church will go back to using a visual canvas to tell the scriptural story of God within their respective worship spaces. Here's why:

First, You may have heard of the term projection mapping before. It's actually very similar to environmental projection in that it uses projection as a way of making blank surfaces seem to have images on or around them. For example, projection onto a building, using the technique of projection mapping, may make the building appear to crumble. It's all about the specific use of quality content, however most projection mapping is custom-designed content for that exact building or project.

Projection mapping is typically not associated with worship as it's often for the sake of entertainment when it's used. The technology in projection mapping and environmental projection are similar, but the intent of use is what separates the two. Environmental projection, when appropriately used, is designed to create an environment that would render you in a posture or state of worship through powerful imagery and collaboration of color and music.

You need to know that projection mapping has been around for a quite awhile. The first [known] projection mapping occurred in 1969, in Disneyland for the opening of the *Haunted Mansion Ride*[1]. That probably shocks most of you, but it's true, projection mapping has been around for several decades.

However, I argue that it dates back even further. The only unique element in projection mapping or environmental projection is the content and the surface on which we're projecting. That means it really isn't that different from the original projectors that surfaced as early as the 1900s. The advancement in modern projection technology and image processing tools have been driving force in making environmental projection more cost effective and accessible.

As you can tell, people have been projecting onto surfaces for a long time. The stories being told through the projected content are different. Using media to take people somewhere completely different or to allow them to experience the immersive mystery of art is different. The idea that with a few thousand dollars and a creative media palette we can literally use media to create an atmosphere—that is what is different!

So no, environmental projection isn't going anywhere, just like music during worship isn't going anywhere. Will it change? Of course. Will it morph? Of course. Because technology is always changing, always innovating and always advancing. However, I stand convinced that technology and cables, projectors and your literal wall aren't what make environmental projection so great... it's the content and the story it tells.

Technology is just the medium. Content is what allows us to take our church body into the cosmos to imagine and contemplate the wonder and majesty of God's creation, or wrap them in the sacred holiness of an ancient cathedral. It's the art we use to paint the walls of our church that ushers in a sense of reflection and curiosity that begs the soul to dig deeper. It's the visuals we use that lead people to see Christ more clearly during Easter as they visualize the pain of the cross and the suffering that took place. Projectors can't do that. Technology isn't what tells the story. Art does, and it's time that art takes center stage.

Consider environmental projection a new paintbrush for your canvas, or a new form of art that allows others to experience the beauty and majesty of visual media instead of just seeing it.

A Cautionary Tale

Technology will always precede artists. As you look over the past several years, you can see this to be the case. The electric guitar came along well before Les Paul was able to give it his signature sound. Portable cameras were around well before Jeremy Cowart and some of the great modern photographers were able to do something incredible with them. Computers existed long before someone was able to connect one to a projector and use it as the vehicle of visual worship. But that artist exists today. It's up to you and I, Bob and Larry.

This means technology will continue to advance. Will it be holographic media? I don't know. Will it be 3D? I don't know that either, but all of this will eventually come. Around every corner we will be asked to proceed or not proceed through the doorway or hallway before us. Be careful not to jump into technology too early and forget the calling to create visual environments that connect people with the ultimate Artist.

As artists, we need to keep the pursuit of telling great stories as our focal point and remember that the technology is just the tool set we use to make stories happen. As technology morphs, so will our use of it.

The Next Generation

I have said time and time again that environmental projection should not. and is not, a way to lure the next generation into your stale or older church family. It's not the way to do a "contemporary" service. Just like you can't simply take off your tie and call yourself modern, you can't put a projector up and call it environmental projection. However, this also doesn't mean we need to alienate the next generation.

You see, there's something beautiful about the circle of life and technology. The next generation will always adapt faster and know newer technology better than previous generations (on the whole). The next generation can be a great resource to us as technology continues to evolve. I am the direct result of the previous generation investing into the next generation. Let me explain.

I wouldn't be here today if it weren't for two guys who reached out to some high school students to help run lighting and slides during the Sunday services. Those two guys allowed me to discover a passion to learn, and a passion to create atmosphere. They invited me into what God was calling them to do for the rest of their lives. However, their decision came with risk.

That risk meant that it probably wouldn't be perfect all the time, because as a teenager I was diagnosed with ADHD (Attention Deficit Hyper-Activity Disorder), which means I probably missed a cue and was eager to over-complicate. However, that was exactly what fueled me to learn, explore and test new things, in a safe environment. These two guys, Matt and Steve, were the control substance in my mad scientist lab.

One key thing to know about these two guys is that they were incredibly passionate about a guy named Jesus and they made it very apparent as we spent late Saturday nights together preparing for Sunday services. We would dig into Scripture, not because we worked at a church, but because we were children of God. This process built for me a foundation that enabled me to leave that church, continue to explore and yet, be grounded in Biblical truth the whole time.

I'm still friends with Matt and Steve and they've played a huge role in my life. What would it look like for us to do the same for a few teenagers in our church today? Would they one day write the book on the next wave of technology for the church? I don't know. I can tell you this, if we don't take a risk at all and we exclude the next generation, the book they write will not be coated in the love of Jesus Christ. They won't identify with Paul in Colossians on the whole "In everything you do, in word and in deed, do it all in the name of the Lord Jesus Christ giving thanks to God the Father."[2]

We can also learn from them. Probably because they have time on their hands and we don't. But mostly because they have a curiosity that is far greater than ours. (That's probably another book for another day.) Learning from the next generation includes them, gives them a sense of belonging, and helps them realize they have a role in the Kingdom of God.

Environmental projection is in many ways the one thing that can reach across the aisle generationally and unify our tech teams. As I mentioned before, the next generation of younger students are eager to get involved. I know this because I was in those same shoes. We must overcome the fear that if we include them we're hurting their ability to learn. If Matt and Steve hadn't reached a hand out and invited me in to help, I wouldn't know God nearly as well as I do today.

Lets begin to include the next generation, not in a way that makes it all about them, but in a way that shows them what it looks like to do media with a heart leaning toward God. The next generation looks to us as the model, which means you and I have to lead and live in a way that our own hearts lean toward God. Let's learn from the next generation and let them teach us throughout the process.

The New Medium

A good friend of mine, and mentor in my life, said to me one night after dinner that if Jesus were to live today in the 21st century, His medium wouldn't be parables, it would be YouTube! If you think about it, it's actually a pretty interesting statement. He went on to describe that Jesus used parables because His audiences weren't very literate. So he painted pictures in verbal form. He told stories that would stick and stand the test of time.

He was using the most viral form of communication around back then, which means if Jesus were around today He would use the most viral form of communication today: YouTube. It's the only platform that, within minutes or a few days, can reach millions of people. Parables were Jesus' form of hoping for a retweet!

Today, we have a new medium in which to tell a story. It's not through the constrained window of a 4:3 screen in your worship center right before the pastor gets up to speak. It's the entire room. It's the entire space. It's environmental projection.

The Journey Continues

When we started this book, I invited you on a journey. Though this book may be coming to an end, the journey continues. Don't let your exploration of this topic stop here. Continue to dig, read, pray, talk it out, try, experiment, get quiet, meditate, try again and read some more. The journey continues for you.

I hope I've been able to expose you to more of the process and philosophy of environmental projection, but there's so much more to learn. What is the visual style in your church? Do you use motions or stills? What is the palette of your community and your body?

There will be a thousand questions to ponder as you embark on this journey, however I ask that you wrestle greatest with one question. Why?

This is the question that will get us to the root of who we are in Christ and what our community believes about visuals in a worship setting. If we just focus on the how and what, we're only going to copy, we'll never create. God doesn't want us to copy. He wants us to create because Christ is most evident in the act of creation not replication. We have to depend on Him more during the creation portion and we're most like Him during the act of creating. It's like the bond a father and son have when they wear the same shirt or like the same things.

Ask deep questions, engage in rich conversation and soak up as much as you possibly can. Continue searching and exploring God's word for imagery, as well. Start with the study of Exodus or begin to look through the lens of creativity and art as you read the Scriptures. It will change the way you communicate with media and it will help you create an atmosphere that beckons people to worship because you will ultimately be leading out of a posture of worship.

Wrestling

Questions will bring doubt and uncertainty. I have come to be okay with questions. It's a weird thing, to be honest. I find that the more questions I ask, the wider my perspective of God becomes. It allows me to be more acutely aware of the areas that are damaging to a visual environment and to be more in tune with what the Spirit is doing in a time of worship.

The more I pursue and learn, the more I doubt. It wouldn't be authentic if I didn't. I wrestle with some key topics that have to do with environmental projection and immersive environments. I question if we as humans are trying to control the environment too much, wrestling with the balance of letting the Spirit move and planning enough to create something powerful.

I wrestle with the idea that some may be distracted while others may be deeply impacted and wonder how to create an environment that is suitable for both. I wrestle with too much of "it" and too few seasons where it is turned off. I wrestle with too many particles. I wrestle with too many crosses. I wrestle to find great media sometimes and I wrestle to pick the one piece of media out of a slew of great pieces that tells the story the way God wants it to be told, not the way my business or brand wants it to be told.

I wrestle with the idea that it can be too bright and at the same time too dark. I wrestle with the idea of environmental projection being noise in an already loud environment. I wrestle with making sure Jesus is the center instead of the wow factor.

I wrestle with so many things because I am still doubting, but I stand convinced of one thing. All worship is art. It's all art because our worship is creating a response to the being of God. Whether it's media, creativity, visuals, environmental projection, music, Sunday, prayer, conversation, or a meaningful conversation with a friend away from church, it's all worship and it reflects the deep, inner desire we have to create (or not create) something that's of worth for our Creator.

Mostly, I wrestle with making sure that whenever I'm behind a computer leading visuals in an environmental projection atmosphere, I step out of the way and allow God to take over. I wrestle with putting my own desires to the side and letting God communicate powerfully and boldly to His people, including me.

I want one thing for you as we continue on this journey, I want your passion for God to outweigh your passion for media, creativity and visuals. I want your art in general to flow out of the abundance of love you have for Jesus Christ. I want you to position yourself in a posture of continual worship and adoration and lead out of that posture. The most amazing worship leaders I've been blessed to have lead me in worship, don't lead out of technical ability or a mental mind game, they lead from the heart.

Lead from the heart.

Explore your God-given gifts.

Create atmospheres.

Change lives!

NOTES

Chapter 1: Nothing New

1. "Cathedral" Wikipedia: The Free Encyclopedia. Wikimedia Foundation, Inc., 26 Feb, 2016. Web. 2 Mar. 2016. <https://en.wikipedia.org/wiki/Cathedral>

2. Hart, Vaughan (1995), St Paul's Cathedral, Christopher Wren., London: Phaidon Press Limited, p. 17–23.

3. "Cathedral Building in the Middle Ages" Durham World Heritage Site. Durham World Heritage Site, Web. 2 Mar. 2016. <https://www.durhamworldheritagesite.com/architecture/cathedral/construction>

4. New International Version. [Colorado Springs]: Biblica, 2011. BibleGateway.com. Web. 2 Mar. 2016.

5. 1 Corinthians Chapter 12, New International Version. [Colorado Springs]: Biblica, 2011. BibleGateway.com. Web. 2 Mar. 2016.

6. C N Trueman "Medieval Cathedrals" The History Learning Site. 5 Mar 2015, Web. 2 Mar 2016. < http://www.historylearningsite.co.uk/medieval-england/medieval-cathedrals/>

7. "Cathedral Building in the Middle Ages" Durham World Heritage Site. Durham World Heritage Site, Web. 2 Mar. 2016. <https://www.durhamworldheritagesite.com/architecture/cathedral/construction>

Chapter 2: Sterile To Sacred

1. "CHARACTERISTICS OF U.S. CONGREGATIONS, BY FAITH GROUP – PART 1" United States Congregational Life Survey. Presbyterian Church U.S.A., 2008. Web. 2 Mar. 2016. <http://www.uscongregations.org/blog/2014/02/17/characteristics-of-u-s-congregations-by-faith-group-part-1-20082009/>

2. "Speakers Give Sound Advice". Syracuse Post Standard (page 18). March 28, 1911.

3. "How Many Words Are There in the Harry Potter Book Series?" Word Counter Blog. 23 Nov 2015. Article. 2 Mar. 2016. <https://wordcounter.net/blog/2015/11/23/10922_how-many-words-harry-potter.html>

4. Trauzettel-Klosinski, Susanne; Dietz, Klaus (August 2012). "Standardized Assessment of Reading Performance: The New International Reading Speed Texts IReST". Investigative Ophthalmology & Visual Science 53 (9). <http://iovs.arvojournals.org/article.aspx?articleid=2166061>

5. The Internet Movie Database. IMDb.com, Inc, 2009. Web. 2 Mar 2016. <http://www.imdb.com/>

6. Hipps, Shane. Flickering Pixels. Grand Rapids: Zondervan, 2009. Book.

Chapter 3: The Manifesto

1. New International Version. [Colorado Springs]: Biblica, 2011. BibleGateway.com. Web. 2 Mar. 2016.

2. McElroy, Luke. The Wide Guide: Blueprint for Multiscreen Media. Nashville: Orange Thread Media, 2014. eBook.

Chapter 10: The Fifth Ingredient

1. Acuff, Jon. Start. Nashville: Ramsey Press, 2013. Book.

Chapter 11: Step By Step

1. Ramsey, Dave. Total Money Makeover: A Proven Plan for Financial Fitness. Nashville: Thomas Nelson, 2007. Book.

2. Website. Crowd Control Games. 2 Mar 2016. <http://CrowdControlGames.com>

Chapter 12: Choosing the Right Media

1. Osgood, Charles E. 1953. Method and Theory in Experimental Psychology. New York: Oxford University Press.

Chapter 13: Parting Thoughts

1. Jones, Brett. "The Illustrated History of Projection Mapping" Projection Mapping Central. 20 Dec 2012. Article. 2 Mar. 2016. <http://projection-mapping.org/the-history-of-projection-mapping/>

2. New International Version. [Colorado Springs]: Biblica, 2011. BibleGateway.com. Web. 2 Mar. 2016.

ABOUT THE AUTHOR

Luke McElroy is the founder of Orange Thread Media, the parent company to TripleWide Media, SALT Conferences and Orange Thread LIVE. Through his leadership, Orange Thread's work has been seen around the world with well known brands including American Idol, Blake Shelton, Bill Engvall and hundreds of churches every week through the stock media their team has curated. Luke was recognized as a leading entrepreneur in his industry by being named one of the top innovators in the church by *Worship Leader Magazine* in 2013 and his inclusion on the *Impact 100* "List of Entrepreneurs to Watch Under 30".

Orange Thread Media is a leader in creating products, resources and events that center around creating engaging environments in a live setting. Previous to this book, Luke is the author of *The Wide Guide: Blueprint for the Multiscreen Movement*. He currently lives in Nashville, TN.

LukeMcElroy.com

ACKNOWLEDGEMENTS

This book would not have come to fruition if it weren't for a group of amazing people.

First, I'd like to thank Tim Southwick, who has championed this book since the day he first heard about it... Thank you for your patience, your insight, your perseverance and commitment to this project, without you it would not exist. Period.

Rachel Dawn Hayes, Thank you for helping me refine my thoughts by providing much needed editorial guidance. Victoria Aebischer, Dan Almond, Taylor Valarik, Teddy Cheek and the rest of the Orange Thread Media team who allowed me space and time to capture these thoughts on paper.

Camron Ware. If it weren't for you being willing to experiment and take risk, Environmental Projection wouldn't be a concept. Thank you for your friendship and being among the best integrators of this concept. Keep being amazing at what you do. I'd also like to thank Stephen Proctor for your friendship and your ability to challenge, equip, inspire and push me in my thoughts on visuals in worship. I'll never forget the "fire pit" where such amazing things were birthed as we (among others) began to trust God for the future of creative arts in the Church.

Steve Simmons, Matthew Reeves and the Mount Pisgah UMC community who supported me when I was getting started with creativity in the church. If it weren't for your early encouragement, I wouldn't have been able to start this creative visual journey.

I'd also like to thank the entire TripleWide Media community for letting us do this. When we started this mission in 2008 of *Fueling the Multiscreen Movement,* we never imagined we would have a customer in almost every country in the world along with producers who are as quality of artists as we have.

Of all the people who I've thanked, the two that have been most instrumental in allowing me to do what I do every day are my mom and dad (Chuck and Lori McElroy). Your encouragement for me to take risk, follow God at every step and do what I love became the foundation for this book and our ministry. I love you both so much!

Finally and most importantly; My Lord and Savior Jesus Christ. There is nothing on earth as important as realizing He died for my sins, allowing me to live in paradise one day in the midst of His presence. It's that passion that drives the desire for people to create powerful visual environments.

BONUS RESOURCES

Since this publication was originally released as a digital product, many aspects of this book are best viewed on an E-reader or computer. In order to help enhance your experience, we're giving you access to the digital version of this book. It is for personal use only, and any unauthorized distribution may be punishable by United States Copyright law.

Also included on this bonus resource page are diagrams, enlarged color photos/pictures, links, free TripleWide Media content and much more!

For more information or to download bonus resources please visit: http://LukeMcElroy.com/EPBook-Bonus

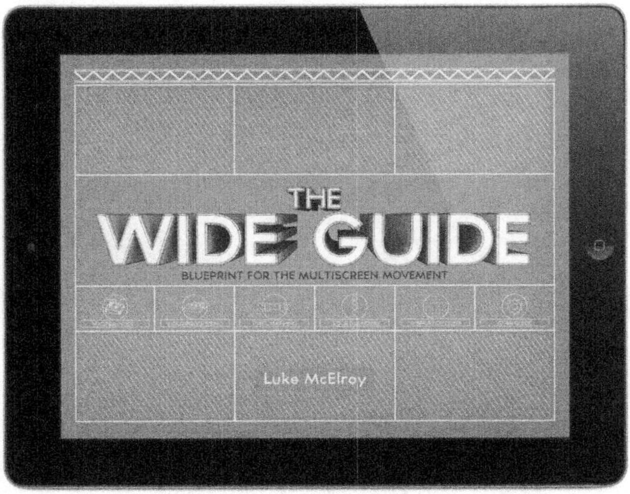

ALSO BY LUKE McELROY

In a generation where technology has never been more affordable, content never more accessible and creativity seemingly endless, The Wide Guide exposes all the secrets to multiscreen video production. Expert and author Luke McElroy, founder of TripleWide Media, beautifully illustrates (literally) all the tips and tricks of setting up multiscreen video displays and shows you the tools needed in order to look and feel like an expert yourself. This is a must have for any production freelancer, creative producer or stage designer.

Available today for Kindle and iBooks

Made in the USA
Monee, IL
04 November 2026

67384999R00085